平面港

GRAPHICS HARBOR

编排设计

成朝晖 编著

中国美术学院出版社

平面港
GRAPHICS HARBOR
编排设计

Contents　目录

"设计师——作为现代社会的艺术家——还必须在其设计的作品中提供丰富的文化营养、情趣和别的什么有价值的东西。"

——[英]艾伦·弗莱彻

"在我所有的广告招贴作品里，首要的——同时也是最重要的——就是关于设计意念的问题。而设计的意念又来自于我所直接面临的设计问题之中，诸如它的本质、背景、对象、受众、功能范围和预算费用，等等。"

——[美]兰尼·索曼斯

"求真是设计的本质。"

——[日]松永真

"我认为设计最重要的不是视觉上的个人风格，而是作品中的神韵。'神'是我多年来在创作中努力追求的东西。"

——[日]松井桂三

一、编排设计概述

编排是以优秀的布局来实现卓越的设计的。

编排设计是现代艺术设计的重要组成部分，也是视觉传达的重要手段。它是将文字、插图、照片、图形、标志等视觉元素进行有机的整理与配置，做总体的安排与布局，将理性思维个性化地展示出来，使其成为具有最大诉求效果的构成技术。成功的编排设计一方面是通过动态、视觉诱导、空白运用、结构、比例等艺术手段，处理各种具有不同作用的构成要素，使之达到均衡、调和的效果，使其成为一个具有视觉魅力和强而有力的组织构成，给消费者提供正确、清晰、完整而明快的信息；另一方面是通过设计师个性化的风格和具有艺术特色的视觉传达，使观者产生感官上的美的享受，并使设计在效果与功能上事半功倍。有人将编排设计师比喻为音乐作曲家，将各种不同色调、肌理与形态的视觉要素组织成为变化丰富而又高度统一的优美乐曲；也有人将其视为舞台中的场景调度，将各种承担信息传达任务的文字、图形艺术化地组合起来，使整体设计变成一个有张有弛、刚柔并济、充满戏剧性的舞台。一份充满动感与节奏感、形态张扬的编排设计与一份严谨、清晰的编排设计传达给人的心理感受是截然相反的。

形式美的法则在编排过程中起着决定作用，它使设计的版面效果生动、简洁、典型、引人注目。而研究和突破形式法则会使编排设计形式独具格调和个性。

编排设计的范围可涉及平面设计的各个领域,如产品简介、企业样宣、海报、挂历、贺卡、包装、报纸、杂志、书籍、画册、信封、信笺、名片、POP 等等,可以说平面设计的原理与理论贯穿于编排设计的始终。

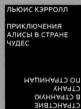 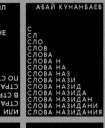    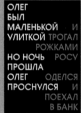

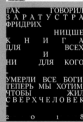    

Typography Program BBE 2.0 — 3.0 编排设计

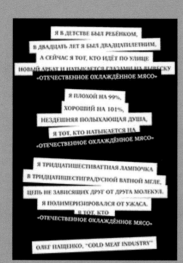

二、编排历史沿革

人类早期文明的发展阶段，无论是在岩洞石壁上的随手涂鸦，还是在泥板、兽骨上刻写的象形文字，就已出现了一定的编排意识。例如两河流域文化时期的锲形文字，书写者已将文字用直式线条分隔，使画面出现一种节奏上的变化，这大概是平面上各要素进行分割处理的最初尝试。古埃及人运用象形文字与插图，书写了反映当时的文化、政治、经济与宗教的文献，从平面设计的角度而言，图文并茂，错落有致，具有较高的设计水准。从此，类似的编排设计一直延续下来，在中国发明印刷术和造纸术以后，得到进一步的发展。

编排设计理论的形成源自 20 世纪的欧洲。英国的设计家在原有的维多利亚式精致、浪漫、复杂的设计风格上发展了一种被称为是工艺美术运动的设计风格，以实用设计家威廉·莫里斯为代表人物。他的设计涉及许多领域。在平面设计中，他尤其讲究版面编排，强调版面的装饰性，通常采取对称结构，形成了严谨、朴素、庄重的风格。莫里斯的古典主义设计风格，开创了编排设计的先河。以他为首的工艺美术运动设计家，创作了许多被以后设计家广泛运用的编排构图方式，如：将文字和曲线花纹结合在一起而形成优美又富有浪漫情调的图文装饰；将各种几何图形插入用于分隔画面；等等。工艺美术运动是世界进入现代工业社会后第一个具有广泛影响的设计运动。尽管他们的设计方法和设计风格十分精美，但过于复杂而显得烦琐，具有制作成本高、工艺技术复杂的缺点。此外，图案繁复造成了某种视觉传达的障碍，不能从根本上解决工业化之后存在于设计与现代工业技术之间的矛盾。

1919 年，随着德国包豪斯设计学院的创立，它所提倡的构成主义运动推进了编排设计的发展。包豪斯设计学院的设计家们根据技术与社会发展的需求，提出了一套完整的设计方法。他们将数学和几何学应用于平面分割，为骨骼法的创

威廉·莫里斯设计集

造奠定了基础；运用几何图形和文字设计的招贴让人们感受到一种全新的视觉设计表现语言；对抽象图形，特别是硬边几何图形在平面设计上的应用进行了全面的研究和探讨；提出了"功能决定形式"的主张，形成了一场席卷全球的现代主义设计运动。另外，俄国的构成主义倡导者李捷斯基提出的"构成主义是理性和逻辑性的艺术"影响了一代设计师。其造型元素出自视觉的基本成分——点、线、面、节奏、色彩、范围、位置以及方向。点、线、面的组合、延伸影响了视觉并引起观者的情绪波动。构成主义的艺术探索适合大工业生产的时代，人们可以接受并认同这"美与正确"的概念。德国设计家约翰借此发展出新客观主义理论，成为编排设计的重要里程碑。其基本原则是彻底脱离传统的版面设计，采取非对称式结构，强调强烈的明暗对比以及块面、粗线条的运用；注重个性化的表现，风格活泼多变；同时还强调编排设计的功能效用，进一步寻求内容与形式之间的有效结合。第二次世界大战的爆发后，大量设计家逃亡至中立国瑞士，同时也将最新的设计思想和技术带到了这个国家，使其在一段时间内走在世界设计发展的前沿。设计家们对骨骼在平面设计中的运用进行了全面研究，形成"瑞士骨骼"的编排方式，影响了许多国家的设计，成为新设计运动的典范。二次大战结束后，随着经济的复苏，许多国家在设计与文化方面迅速发展。其中日本尤为突出，其设计师广泛吸收自己需要的有益养分，探索出极具现代性同时又具有日本传统色彩的象征性和简洁性的风格：在编排方面强调大胆的分割，以平面化、简洁、凝练的处理取胜；注重余白，追求平淡内敛的阴柔美的意境，赢得世界设计界的瞩目。20世纪七八十年代至九十年代，随着电脑技术的发展与普及，互联网的出现意味着更多的设计表现语言和制作手段的出现。新的编排形式的出现——图形的多样化的组合、多层次的复叠、空间的延展、画面的肌理等，使编排设计语言更为奇特绚丽。设计师日益突破传统框架的限制，比以往更为自由地驰骋想象。

包豪斯风格印刷字体

赫伯特·巴耶的封面设计

包豪斯设计学院的展览海报设计

包豪斯设计学院的芭蕾舞表演海报

三、编排设计价值

编排设计是依照视觉信息的既有要素与媒体介质要素进行的一种组织构造性设计：根据文字、图像、图形、符号、色彩、尺度、空间等元素和特定的信息需要，按照美感原则和人的视认阅读特性进行组织、构成和排版，使版面具有一定的视觉美感，适合阅读习惯，引起人的阅读兴趣。

版面编排设计的最终目的在于使内容清晰、有条理、主次分明，且具有一定的逻辑性，以促使视觉信息得到快速、准确、清晰的表达和传播。对一个特定信息界面的具体编排而言，各种元素的统筹和富有创意的表现，不仅是方便阅读的需要，也是产生视觉美感的需要，因此形式美的法则是影响版面编排优劣的决定性因素。

符合形式美法则的编排设计能使版面简洁、生动、充实、协调，更能体现秩序感，从而获得更好的视觉效果。

版面的形式美感可以概括为几种形式：明快统一、井然有序的对称形式感；富于变化、生动的均衡形式感；重复统一的节奏形式感；轻松、优雅的韵律形式感；简洁、对比的空间形式感；生动、富有创意的变异形式感；并列、清晰的网格形式感；整体、协调的统一形式感。

LIFET 系列
设计：永井一正
主海报文字在视觉中心的附近，突出海报的核心主题。

此编排一反传统的美学规律，采取极端的自由无序，形成互相冲撞、互相消耗、互相矛盾的态势，以达到更为耗散发射的视觉效果。

四、编排设计原则

编排设计是一种形式语言探求活动，但它的孕育原点是与所要表达的内容特征紧密匹配在一起的。所有的技巧都旨在清晰、鲜明地传达内容要素，用悦目的组织排列来更好地突出主题，达成最佳的诉求效果，而非为了形式而形式，或者为了形式而寻找内容，这些是编排设计的原则。

1. 鲜明突出地诉求主题

主题的突出可以通过编排的空间层次、主从关系、视觉流程以及对彼此之间的逻辑条理性的把握与运用来达到。编排是为了构成所表述之内容活性化的流动之美，追求合理形式以符合主题的思想内容，以鲜明、独特的形式美为清晰、准确地传达丰富多彩的内容服务，这是编排设计的前提。脱离内容的形式，或者只求内容而缺乏艺术形式的表现，都会显得设计空洞、乏味，也无编排设计的意义。只有内容和形式的完美结合，才能共同创造出形神兼备的、具有生命力和保存价值的设计作品。

TONG KUNNIAO 编排
设计：David Rindlisbacher、Maximilian Mauracher
大标题字体与说明文字组合，形成简洁且严谨的排版。

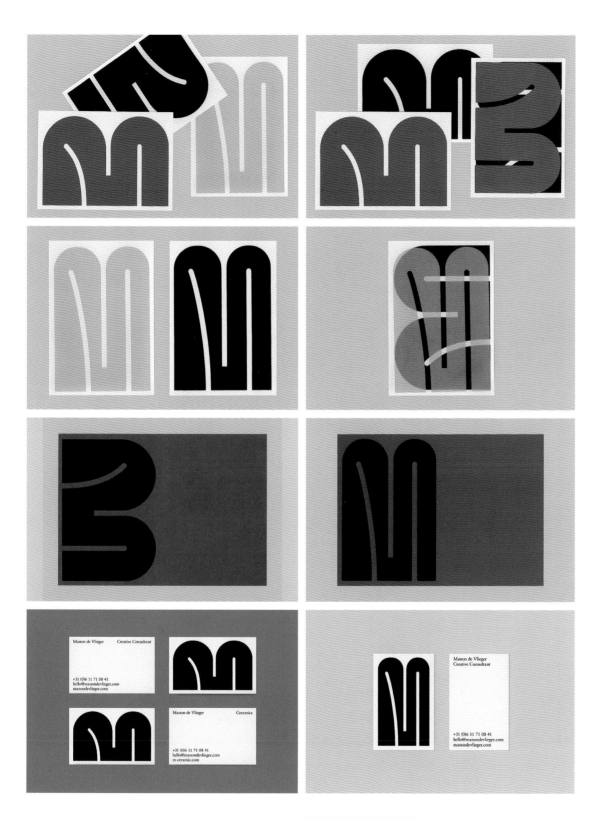

MANON DE VLIEGER 识别
设计：Studio Lennarts & De Bruijn
以 M 为重点，各种色彩与 M 的黑色搭配，创造一个清晰而简单的标志和身份。

TONG KUNNIAO 编排
设计：David Rindlisbacher、Maximilian Mauracher
箭头成为贯穿每个页面的形象，形成自由和有导向性的引导。

2. 在统一之中求得变化

在统一变化之中求得，即既有变化又有统一。通过大小、形状、方向、明暗、动静以及情感变化等对比，利用空间对比、虚实对比、平面与立体对比、色彩变化等，能使得版面更加悦目生动。缺乏对比的编排设计，往往缺少视觉冲击力。而有比较、有差异的编排才能引人注目。广告的各要素之间要具有一种共通性，使有所差异的各要素统一协调，成为一个整体，使内容的表现更有创意、更有感染力，以达到最大的诉求效果。有机地保持"变化"与"统一"，呈现充满生机的状态，才能获得高层次的审美快感，取得丰富而又协调的良好画面。

3. 组织严谨且具视觉魅力

等待设计师进行创作的一个平面，就如一张白纸，是有生命的。设计师是调动自己所有的心智、感情与想象力，将各种文字与图形按照视觉美感与内容上的逻辑统一起来，使之形成一个具有视觉魅力和组织严谨的"生命体"，将特定的信息清晰、迅速地传达给每一双掠过它的眼睛。

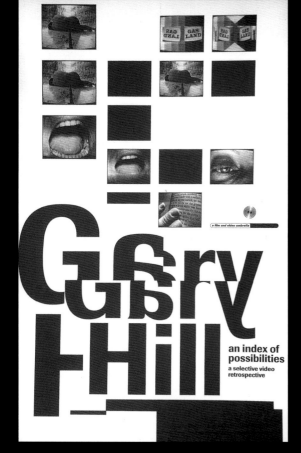

Gary Hill

**an index of
possibilities**

a selective video
retrospective

这两幅作品在编排方面强调大胆的分割，以平面化的简洁和凝练的处理取胜。

The
Modern
Poster

A REVIEW OF CONTEMPORARY POSTER DESIGN BY MICHAEL JOHNSON, JOHNSON BANKS

圆形的文本规格，文字和图形形成圆形的围合，或发射的排版，文字大胆且张扬。

五、编排图文要素

编排设计是在有限的版面空间里，将文字、图形、标志、线条和色彩等版面的构成因素，根据特定的内容进行排列和组合，通过文字的数量、面积、方向、位置等因素产生新的版式编排变化，并运用造型要素及形式原理，把构思与计划以视觉形式表达出来。寻求艺术手段来正确地表现版面信息，是一种直觉性、创造性的活动。因此，设计师要整体地进行思考，尽量将画面组织成一个既具有内在丰富的层次与变化，又浑然一体，具有明显的风格特征的整体。

1. 点

编排中有形。这里所指的形，是在视觉上形成有一定辨识度的形态，在画面上可以有大小、色彩、肌理和外形的变化。形是一切视觉要素的基础，点、线、面都是一种形。

点是最基本的形。在编排设计中，点是相对而言的，而且必须是指可视的形。它既可以是一个形态，也可以是一块色彩，甚至是一张小小的图像。需要指出的是，文字也可以是一个点。

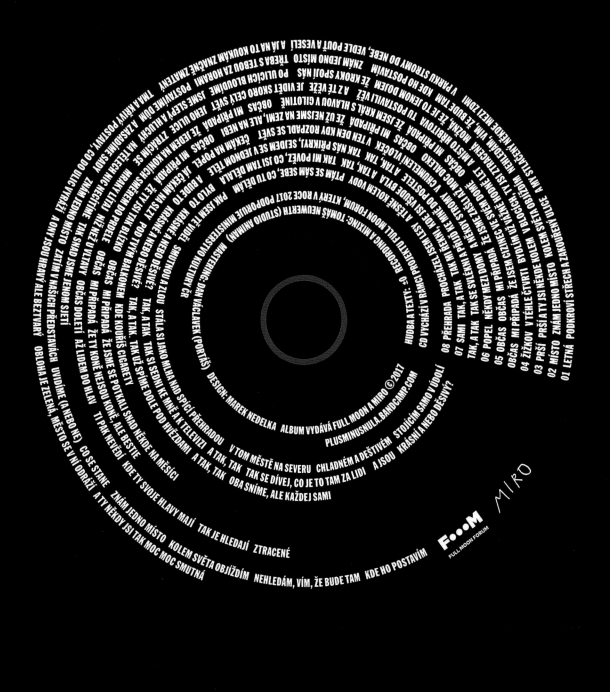

Nedelka 创意指导和平面设计

2. 线

线是点的发展和延伸。线是有性格的形，它有形状、色彩、肌理等多种变化。比如，钢笔画出的线和毛笔画出的线具有完全不同的性格，硬边的线和柔边的线也给人截然不同的视觉感受。同时，线还可以表达出静态和动态，具有长短、粗细、深浅、正负等变化。

文字构成的线，往往占据着画面的主要位置，成为设计者要处理的主要对象。线也可以构成各种装饰要素，以及各种形态的外轮廓。它们起着界定、分割画面各种形象的作用。线既可以有形，也可以无形。如编排设计中的轴线、骨骼线等，它们以一种"无形"的方式达到组合、整理各种图像、文字的作用。线也可以生长、延伸，分割画面。

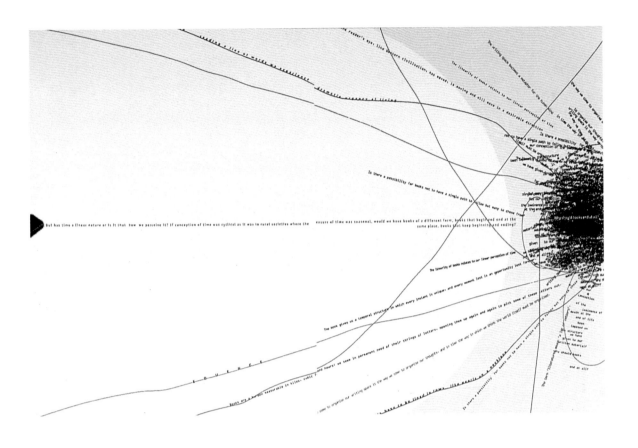

3. 面

面是线的发展和延续，是各种基本形态中最富于变化的视觉要素。面包括各种色彩、肌理等方面的变化，同时，面的形状和边缘对其本身的特质也存在很大的影响。在编排设计中，一个将被设计的画面本身也是一个面。

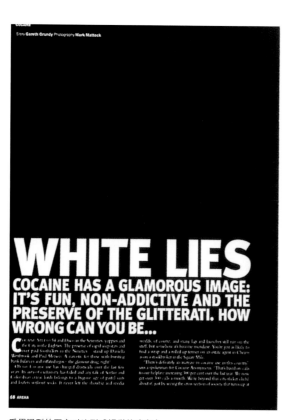

采用强烈的图文对比形成视觉的冲击力。

GREY BOY FROM SOFT CLASSROOM CHARCOAL TO SPACE-AGE SILVER: TONES OF GREY PUT BLACK IN THE SHADE

WHETHER AS ACTOR, BOXER, HELLRAISER OR

HUSBAND, **MICKEY ROURKE** HAS

ALWAYS CAST HIMSELF AS THE OUTSIDER.

THESE DAYS, HOWEVER, THE NEW, REFORMED,

MELLOW MICKEY IS BACK AND THIS TIME

CARROT-TOP WILL BE STANDING TALLER MAN.

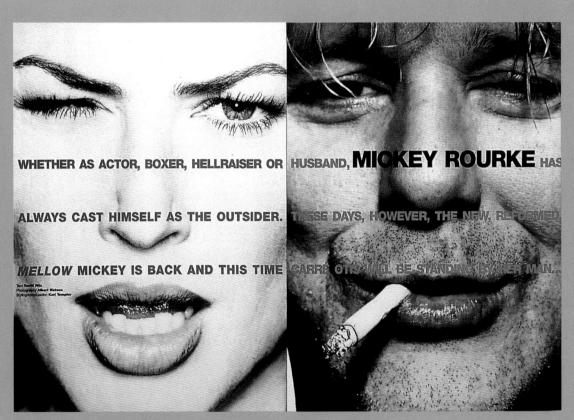

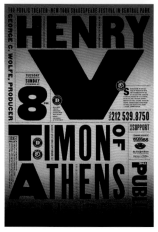

The Public Theater
设计：Pentagram 团队

4. 色彩

色彩有色相、艳度、明度等方面的基本性质。色相是指色彩的相貌，是区别色彩种类的名称；艳度是指色彩的纯净程度，也是色彩的饱和度；明度是指色彩的明暗程度，也可称之为色彩的亮度或者深浅程度。在色相间的各种关系中，冷色和暖色两个组群的关系是最为主要的关系，色彩的冷暖常常是画面的主要表现要素。关于艳度，高艳度与低艳度的色彩可以给画面完全不同的视觉表现，运用不同艳度的色彩，则是调节画面色彩关系的重要手段。在编排设计中，常用高调、中调和低调来概括各种色彩明度的分类。

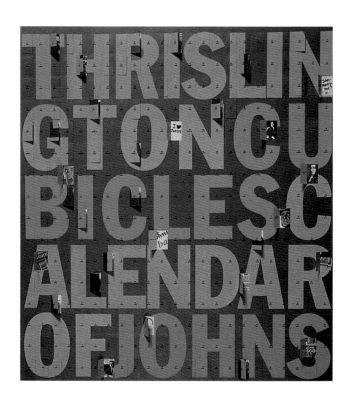

Women's Foundation 25th Anniversary 妇女基金会 25 周年纪念日
设计：Morgan Stephens

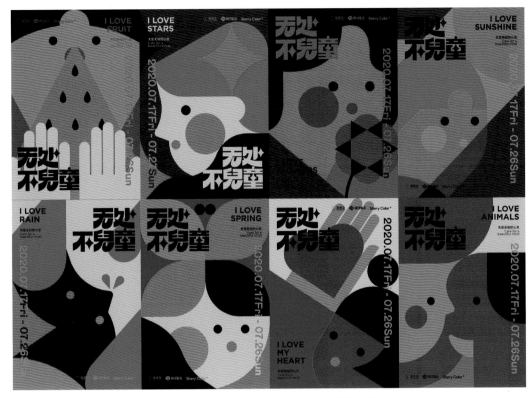

"无处不儿童"展览海报设计

设计：Cheng Peng

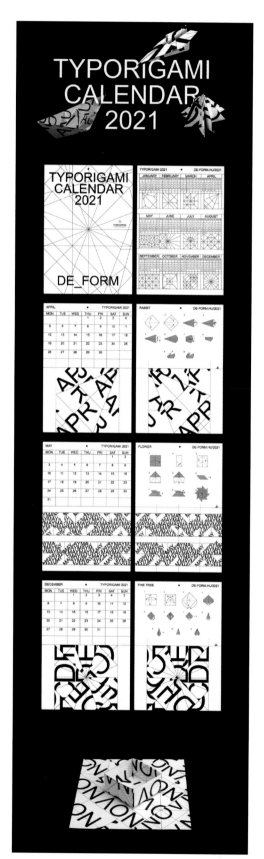

5. 群组

从编排设计的角度来讲，只要设计中出现了各种要素，那么这些要素之间的首要关系便是群组，编排设计的核心问题就是处理好群组问题。群组包含编排的组合和层次。

组合是对画面中部分视觉要素之间关系的组织与处理。它包括：要素在视觉强度上的组合，如疏与密、强与弱；空间关系上的组合，如均衡、对称；形态造型上的组合，如连接、复叠、透叠等；色彩、肌理方面的组合，如调和、呼应等；综合性的要素组合，如对比、统一等等。对于字体组合，运用轴线是很重要的一种方法，它可以使各种字体的结合变得更加有序。轴线既可以是无形的，也可以成为一种装饰线条。理性的编排设计主要运用到骨骼设计，进行骨骼设计前要先根据内容和设计风格的要求设计或者选择骨骼。

将画面分为多个层次，可以更加有效地分析和组织各种复杂的要素。层次是通过形的特性，包括大小、多少，色彩的色相、明度、艳度，以及不同的肌理，空间的位置、动势的抑扬等多种要素来实现的。在设计中，我们既可以将图像图形分层次，也可以将字体分成不同的层次。视觉密度的控制、色调的高低处理、各种肌理的对比都是层次要处理的问题。再从整个画面、整个设计的角度来思考和处理各部分要素之间的层次关系，是对整体性、艺术风格以及效果的把握。

2021 里加米日历
设计：de_form studio
字体组合游戏的方式形成日历的排版节奏。

感恩而死乐队唱片
以歌词和相关图像进行现代的重新诠释,在整个排版设计中使用了油和水的纹理和手绘插图,数字则成为串联编排的要素。

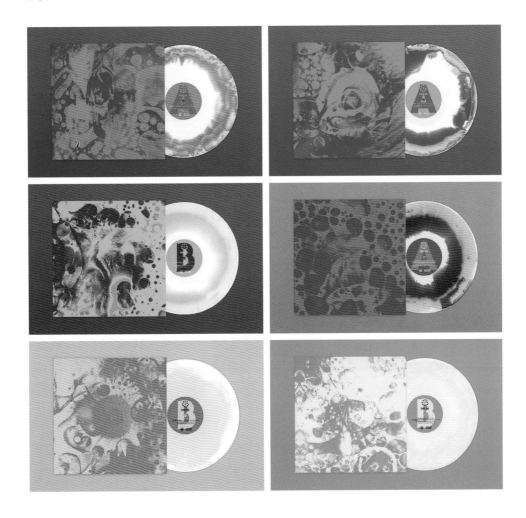

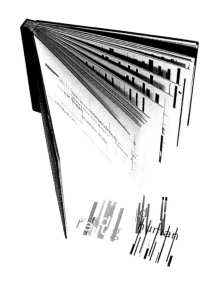

6. 动势

编排设计中的动势主要分为三种：动力、张力和重力。

动力是指平面中自身造型向一定方向伸展流动所产生的视觉力量；张力是形体向空间发散、扩散的视觉力量；重力是由于形体肌理色彩的不同产生的轻重、前后的视觉力量。心理的联想、视觉的冲击都可以造成动势，观者的视觉流程是一种流动的动态过程，画面会形成一种有节奏变化的动势组合。

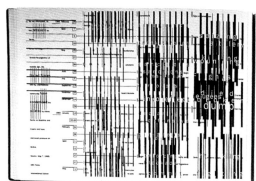
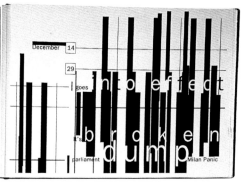

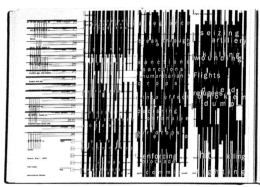
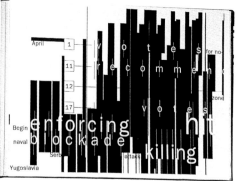

以相似的直式线条与字母穿插排列，形成富有动力、张力和形式感的有规律的统一排版。

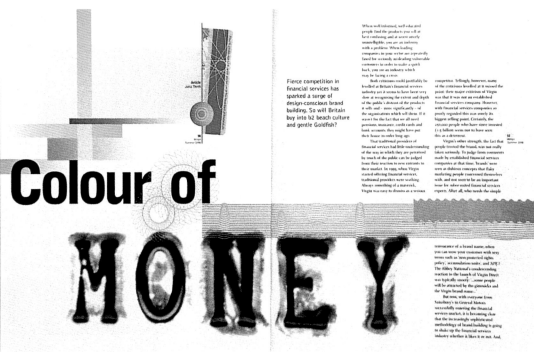

Article
Julia Thrift

Fierce competition in
financial services has
sparked a surge of
design-conscious brand
building. So will Britain
buy into b2 beach culture
and gentle Goldfish?

Colour of
MONEY

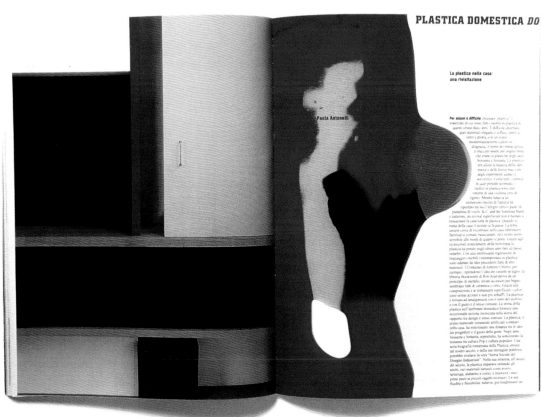

PLASTICA DOMESTICA DO

La plastica nella casa:
una rivisitazione

Paola Antonelli

脱离传统的编排设计，采取非对称式结构，强调强烈的明暗
对比以及块面、粗线条的运用，注重个性化的表现，风格活
泼多变，寻求内容与形式之间的有效结合。

27 [编排设计]

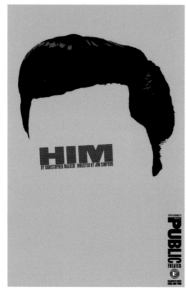

六、编排设计语法

　　作为视觉传达设计，编排必须做到所要传达的信息在逻辑上的一致性、条理性与合理性。要想使广告效果引人注目，其形式美的法则在编排过程中起着决定性的作用，它使版面效果简洁、典型、生动。研究和突破形式法则则更会使编排设计形式独具格调和个性。编排设计形式法则要点如下：

1. 视觉中心说

　　眼睛的错视、生理机能以及视觉习惯等因素，决定了人们会关注画面中最受注目的地方，这就是所谓的视觉中心。设计师应考虑将重要信息或视觉流程的停留点安排在注目价值高的最佳视域，使得主题一目了然。在版面之中，不同的视域，注目程度不同，心理感受也不同。设计师可根据具体元素的传达需要，依照画面的重心关系，安排元素各自适当的位置。通常，版面的上部比下部引人注目。上部给人以轻松、愉快、积极、扬升之感；下部给人以下坠、压抑、沉重、消沉、稳定之感；左侧一般比右侧注目性高。左侧感觉舒展、轻便、自由、富有动感；右侧局限、拘谨、拥挤、紧凑而又稳重。

The Public Theater
设计：Pentagram 团队
满版的文字与图形的排列，营造饱和的画面信息。

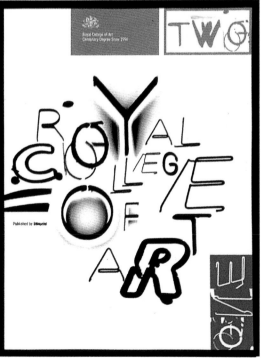

Rolling Stone

Images of Rock & Roll

cyclops
albert
watson

以相同或相似的序列反复排列，形成形象的连续性、再现性和统一性，给人以安定、整齐、秩序和规律的统一感。

2. 视觉流程说

编排设计的视觉流程是一种"空间的运动"，是视线随着各视觉元素在空间沿一定轨迹运动的过程。利用视觉移动规律，通过设计的合理安排，诱导观者的视觉随着编排中有序组织的各要素，从主要内容依次观看下去，如此能使观者有一个清晰、迅速、流畅的信息接受过程。而这个视觉在空间中的流动线是"虚线"，故常为设计者所忽略。有经验的设计师却非常注重运用这一贯穿版面的主线。编排设计只有依靠各个组成部分的相互有机联系，才能渲染出与设计内容相结合的整体阅读环境。这就要求设计师纯熟把握设计中的结构特性和表现特征，掌握彼此之间的和谐关系，吃透它的核心，并由核心向外扩展到各元素的方方面面，然后将与设计的内容相关联的各类要素重新整理、排列、组合、渗透……有节奏、有章法、有创造性地谱写设计的新形象。如果编排的脉络清晰，整体的编排运动有主体旋律，就能更好地引导观者的视线。视觉流程可以从理性与感性、方向性与耗散流程来分析，大致可分为以下几种形式：

A. 单向视觉流程

单向视觉流程往往按照常规的视觉流程规律，诱导观者的视觉随着编排中的有序组织的各元素，从主要内容开始依次观看下去，使版面的视觉流动线具有更为简洁、有力的效果。

直式视觉流程——具有稳定性，是一种强固的构图。视线依直式的中轴线上下移动，给人以直观、坚定之感。

横式视觉流程——安宁而平静的构图。视线会依横式的水平线左右移动，给人以稳定、恬静、平和之感。

斜式视觉流程——是一种强固而有动态的构图。视线一般从左上角向右下角移动，或从左下角向右上角移动，给人以强烈的运动冲击力。

NO GUTS NO GLORY

no shame zone
no shame zone
no shame zone
no shame zone
no shame zone
no shame zone

NO SHAME ZONE

force of nature

START YOUR OWN SHIT

SELF-LOVE IS RESISTANCE

SELF-LOVE IS RESISTANCE

GIRLBOSS GIRLBOSS

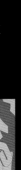

START YOUR OWN SHIT

GIRLBOSS

GIRLBOSS

DO IT YOUR-SELF

force of nature

girlboss moment

GIRL BOSS MOMENT
GIRL BOSS MOMENT
GIRL BOSS MOMENT

girlboss moment

unfuckwithable
unfuckwithable

FORCE OF NATURE
FORCE OF NATURE
FORCE OF NATURE

CHOOSE YOUR OWN GODDAMN ADVENTURE

NO GUTS NO GLORY NO GUTS NO GUTS NO GLORY

GirlbossRally
设计：Kastor & Pollux
一组旨在赋予女性权力的
海报。采用鲜艳的色彩和
大胆的排版来营造一种力
量感和能量感

UNFUCKWITHABLE
UNFUCKWITHABLE
UNFUCKWITHABLE
UNFUCKWITHABLE
UNFUCKWITHABLE
UNFUCKWITHABLE

45°23'10.2"N 1°06'27.1"E 🔍

The "Golden d'Altitude" apple is produced here: **45°23'10.2"N 1°06'27.1"E**. Guaranteed 100% natural, it undergoes no treatment after harvest and saves you from GMOs, preservatives and additives. **And for those who only believe in what they see, just enter these GPS coordinates in your navigation toolbar or scan the QR code to check by yourself.**

Auchan

43°18'21.8"N 5°35'08.9"E 🔍

The zucchini from the southeast is produced here: **43°18'21.8"N 5°35'08.9"E**. Harvested daily for better freshness, it is guaranteed GMO-free. **And for those who only believe in what they see, just enter these GPS coordinates in your navigation toolbar or scan the QR code to check by yourself.**

Auchan

48°08'40.3"N 7°10'55.3"E 🔍

The "Petit Munster" from the Haxaire cheese factory is produced here: **48°08'40.3"N 7°10'55.3"E**. This cheese factory favors short circuits and a reasonable agriculture. **And for those who only believe in what they see, just enter these GPS coordinates in your navigation toolbar or scan the QR code to check by yourself.**

Auchan

49°45'40.1"N 0°22'11.4"E 🔍

The cod is produced here: **49°45'40.1"N 0°22'11.4"E**. There is complete traceability between fishing on the vessel and your plate. **And for those who only believe in what they see, just enter these GPS coordinates in your navigation toolbar or scan the QR code to check by yourself.**

Auchan

编

简练单纯的上下居中的编排设计。
视线会依直式的上下垂直移动，
给人直观、坚定之感。

2.14 GODIVA

安宁而平静的水平式编排形式。
视线会依横式的水平线左右移动，
给人以稳定、恬静、平和之感。

2.14 GODIVA

排版上突破框架的限制，给人以更为自由的驰骋想象！

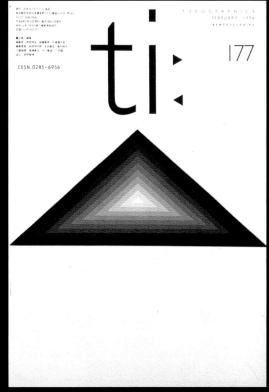

注重余白、追求平淡内敛的意境。

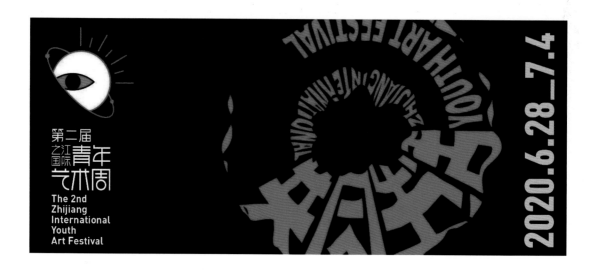

B. 曲线视觉流程

曲线视觉流程是各视觉要素随弧线或回旋线而运动变化的视觉流动。曲线的视觉流程比单向视觉流程更具有明显的节奏与韵律之美，微妙且复杂。

圆形视觉流程——视线依圆环状迂回于画面，可长久地吸引观者注意力，给人以饱满、扩张之感。

S形视觉流程——可将相反的条件相对统一，在平面中增加深度和动感，所构成的回旋富于变化。

C. 反复视觉流程

反复视觉流程是以相同或相似的序列重复，做有规律、有秩序、有节奏的逐次的运动。其运动流程并非像单向视觉流程那样具有强烈的视觉冲击，但更富于韵律和秩序之美。

重复视觉流程——以相同或相似的序列反复排列，形成形象的连续性、再现性和统一性，给人以安定感、整齐感、秩序化和规律的统一感。

特异视觉流程——构成要素在有秩序的关系里，有意违反秩序，使个别的要素显得突出，以打破规律性。由于这种局部的、少量的突变，突破了常规的单调与雷同性，成为版面的趣味中心，产生醒目、生动感人的视觉效果。

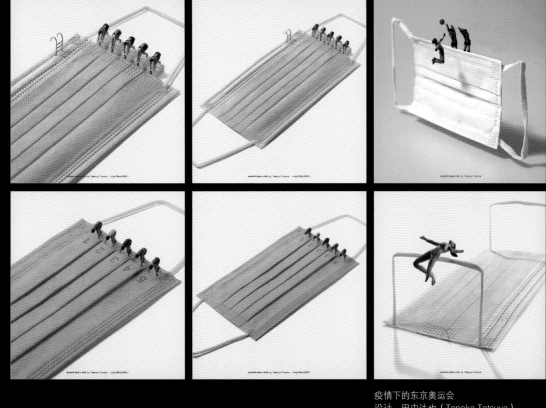

疫情下的东京奥运会
设计：田中达也（Tanaka Tatsuya）
以日常生活中的一些物品作为场景，将一个
个神态各异的小人偶融入，形成一个幽默的
故事或是逗趣的情景。图文采用斜式的视觉
流程排列，单纯强劲。

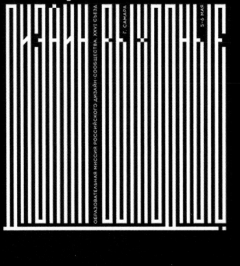

XXVI СЪЕЗД
ДИЗАЙН·ВЫХОДНЫХ

САМАРА
5-6 МАЯ
ХОЛИДЕЙ ИНН

XXVI СЪЕЗД
ДИЗАЙН·ВЫХОДНЫХ

САМАРА
5-6 МАЯ
ХОЛИДЕЙ ИНН

DESIGNWEEKEND.RU

DESIGNWEEKEND.RU

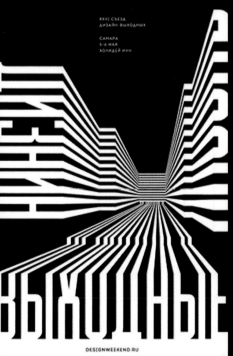

XXVI СЪЕЗД
ДИЗАЙН·ВЫХОДНЫХ

САМАРА
5-6 МАЯ
ХОЛИДЕЙ ИНН

XXVI СЪЕЗД
ДИЗАЙН·ВЫХОДНЫХ

САМАРА
5-6 МАЯ
ХОЛИДЕЙ ИНН

DESIGNWEEKEND.RU

DESIGNWEEKEND.RU

以文字或点、线等形象作为视线诱导的发射式
视觉流程。

排版上突破框架的限制，打破常规的秩序与规
律，以强调感性、随意性、自由性的耗散形式，
给人以更为自由的驰骋想象！

D. 导向视觉流程

导向视觉流程是通过诱导元素，主动引导读者视线向一个方向运动，从而把版面串联起来，形成一个有机的整体，达到多样的统一。

十字形视觉流程——是垂直线和水平线对称的交叉构图，也有斜形交叉的形式。其视线的主眼点集中于十字的交叉点，使得重点突出，发挥最大的信息传达功能。

发射式视觉流程——是以文字或点、线等形象作为视线诱导，将多种条件集中于一个主眼点上，具有多样统一的综合视觉效果。编排中的导线表现多样，虚实相生，富于运动变化。

E. 耗散视觉流程

耗散视觉流程是指版面的图与文字呈自由分散状态的编排，是打破常规的秩序与规律，以强调感性、随意、自由的表现形式。这种编排一反传统的唯美的美学规律，采取极端的自由无序，形成互相冲撞、互相消耗、互相矛盾的态势，以所谓的"混乱"代替明晰，以所谓的"破碎"代替统一，用文字或图像的刻意"缺陷"向结构的稳定性挑战，使设计有悖于理性原则，充满矛盾及偶然的解体形式。这类编排设计往往别出心裁，字体的造型和编排已超越了字符的本体含义，成为作品形象和创意的重要元素，骤看之下近乎"散乱无序"，"有序"为"无序"替代，"完美"为"破败"替代，点、线、面和色块的构成潇洒随意；细看则会发现在富于变化的活泼形象之中，形式关系各要素之间的结构却极为严格，从而在新的变化统一的基础上追求更为新颖、更为潇洒的编排形式。

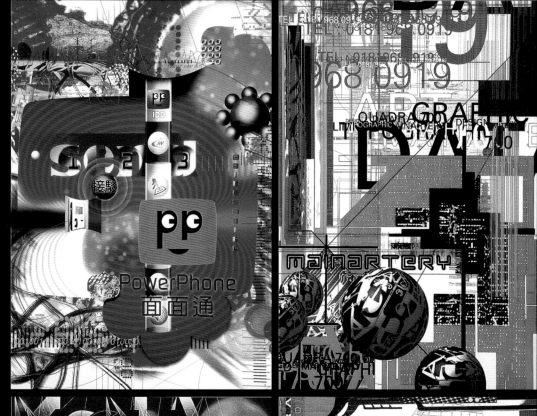

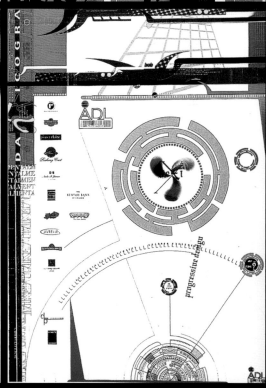

以"无序"代替"有序",以"破败"代替"完美",
点、线、面和色块构成潇洒随意。

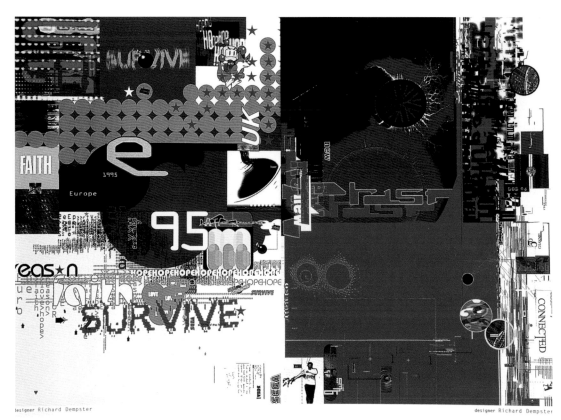

designer Richard Dempster

designer Richard Dempster

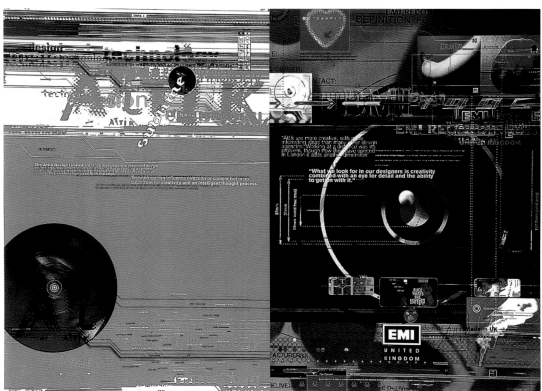

用文字或图像的刻意"缺陷"向结构的稳定性挑战，使设计有悖于
理性原则，呈现充满矛盾及偶然的解体形式。

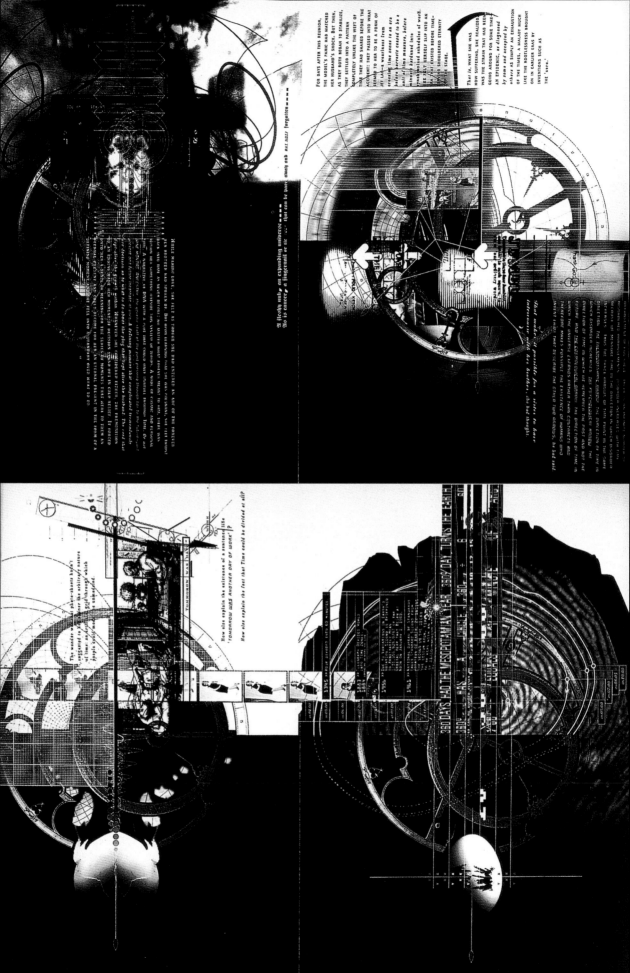

FOR DAYS AFTER THIS REUNION, THE MODEL'S PANIC HAD MATCHED HER HUSBAND'S SHOCK. BUT THEN, AS THEY BOTH BEGAN TO STABILIZE, THEY SETTLED INTO A PATTERN COMPLETELY UNLIKE THE WEFT OF TIME THEY HAD SHARED BEFORE THE ACCOUNT: THEY PASSED INTO WHAT SEEMED TO HER TO BE A FORM OF...

How else explain the subtlety of a sentence like 'TOMORROW WAS ANOTHER DAY OF WORK'?

How else explain the fact that Time could be divided at all?

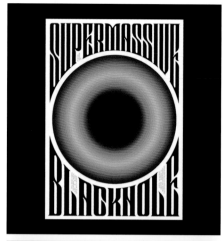

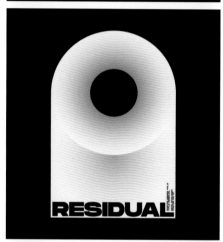

3. 对称与均衡

对称意指各要素以某一点为中心，取得左右或者上下同等、同量、同形的平衡。对称的形式有以中轴线为轴心的左右对称，以水平线为基准的上下对称和以中心为基准的放射对称，还有以对称面出发的反转形式。其特点是明快统一、明确坚实、庄严沉静、井然有序、严谨庄重、高贵、有可信赖之感，但处理不好易单调、呆板。

均衡意指各要素以某一点为中心，在左右或者上下有同量不同形的安排，是与对称相反的"非对称"，然而在造型上，却可取得一种所有要素在整体上分布平衡的状态，这就是均衡。与对称相比，均衡显得比较感性，比较难以理解与把握。如何在感性中总结理性规律，在编排设计中进行实际运用？设计者可将文字图形依形态大小、多少、色调与肌理的明暗等因素，按照各要素主次、强弱的差异关系来设计，取得设计视觉上的美感以及与内容相配的逻辑性。在设计编排表现中，均衡是一种比较自由的重复形式，有时正因为对称反而使版面产生拘谨生硬的感觉，有时又因为井然有序引起单调平凡的感觉。在这种情况下，均衡就变得非常有必要。均衡以等量非等形的方式来表现矛盾的统一性，也是一种在不平衡中求得平衡的设计方法。其目的在于揭示内在的、含蓄的秩序和平衡，达到一种静中有动、动中求静、动静结合的条理之美和动态之美。均衡的形式富于变化和趣味性，具有灵巧、生动、活泼、明快的视觉效果。

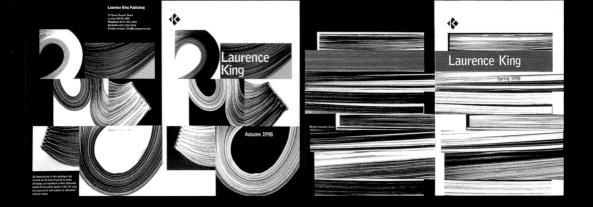

平面设计作品
设计：Studio Mut
对称排版庄严沉静、井然有序、严谨庄重、高贵、有可信赖之感，文字和色彩渐变的处理打破了画面的单调与呆板，简约且智慧。

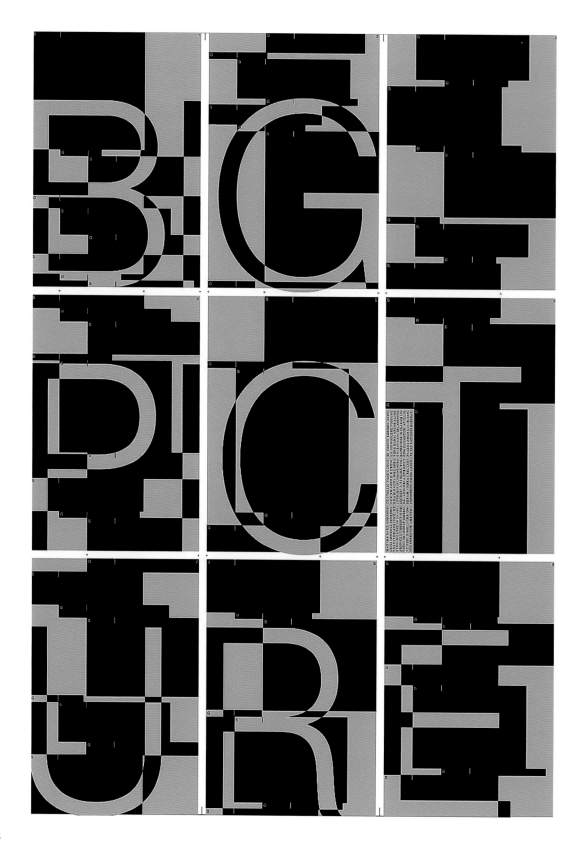

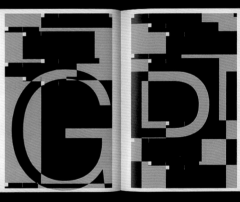

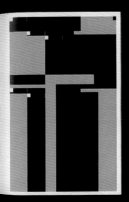

大画面图形叙事
设计：David Rindlisbacher
一本书、一个模拟和一个数字增强现实（AR）海报被特别编排
使得不同的现实感知在多个平台上相互关联。

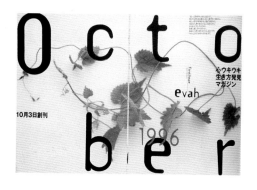

4. 节奏与韵律

节奏与韵律的概念来自于音乐，是体现形式美的一种形式。有节奏的变化才有韵律之美。

用反复、对应等形式把各种变化因素加以组织，构成前后连贯的有序整体即节奏。节奏是艺术表现的重要原则，各种艺术形式都离不开节奏。节奏是按一定的条理、秩序重复连续地排列，形成一种律动的形式。它有等距离的连续，也有渐大、渐小、渐长、渐短、渐高、渐低、渐明、渐暗等排列构成，就如同春、夏、秋、冬的四季循环。节奏的重复使单纯的更为单纯，统一的更为统一。节奏是视觉流程的展开表现，是主观视觉在对画面各种要素观察、读解的流程中，对各种要素间相互关系的一种把握。

韵律不是简单地重复，而是比节奏更高一级的律动，是在节奏的基础上更超于线形的起伏、流畅与和谐。韵律是宇宙之间普遍存在的一种美感形式，它就像音乐中的旋律，不但有节奏，更有情调。它能增强版面的感染力，开拓艺术表现力，牵动人们的思想感情，引起观者的共鸣。

节奏与韵律的运用，能创造出形象鲜明、形式独特的视觉效果，表现轻松、优雅的情感，通过跃动提高诉求力度。

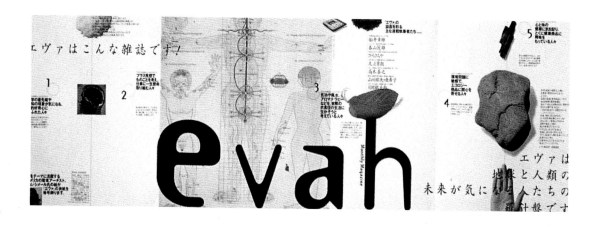

麦当劳广告欣赏：世界自行车日
简洁的图文编排，使视觉流程有轻重缓急地展开表现，

编排

5. 比例与权衡

　　在设计编排上，所谓的比例是量的比率，即长度或面积等的比率。比例是整体与局部以及部分与部分之间数量的一种比率，又是一种用几何语言和数比词汇来表现现代生活和现代科学技术的抽象艺术形式。编排的各要素，以及它们同背景之间形成的比例关系，是形成设计明确的性格特征的手段之一。人们认为的最美比例是黄金分割比率1∶1.618，另外还有等差数列分割、等比数列分割等。视觉比例中，有的好看，有的不好看，根据人的视觉习惯、视觉经验和形式美的原则来衡量、权衡。

　　在美学中，最经典的比例分配莫过于黄金分割了。正如同它的名字，黄金比例常常给人一种恰到好处的自然的感觉，在设计当中被广泛地运用。比例美是人们的视觉感受，又符合一定的数学关系，利用比例完成编排，通常具有秩序、明朗的特征，给人一种清新之感。在排版设计中灵活地运用黄金比例，能够带来不错的效果，可创造出更加聚焦、富有凝聚力的布局。注重比例关系的运用，能使版面编排设计达到和谐、匀称、活泼的美感。

由于局部的、少量的特异突变，突破了常规的单调与雷同性，成为版面的趣味中心，产生醒目、生动感人的视觉效果。

Union des Librairies Musicales 音乐组织联盟视觉形象设计
设计：Brand Brothers

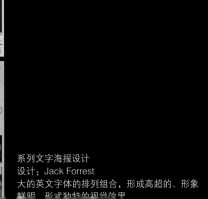

系列文字海报设计
设计：Jack Forrest
大的英文字体的排列组合，形成高超的、形象
鲜明、形式独特的视觉效果。

6. 虚实与留白

空间与形体互相依存。任何形体的存在都占有一定的空间，是实体的空间；形体之外或在形体的背后，或细弱的文字、图形与色彩，就是虚的空间。虚的空间往往是为了强调主体，有意将其他部分削弱，甚至留白来衬托主体的实。空间是编排设计的主要关注点。空间的不同部分对观察者的视觉吸引力是不同的。在设计中，虚的空间有时比实的空间更加重要。虚实之间的处理应给予相应的容纳空间，实与虚的对比是在一个既定空间里配置要素的主要方法。因此，留白是编排设计中一种特殊的手法。实体的空间和虚的空间之间没有绝对的分界，每一个形体在占据一定的实体空间后，还需要一定的虚的空间，使其在视觉上的动态与张力得以延伸。留白对于突出主体、创造意境有重要作用。中国画中向来重视空白的运用，传统美学有"形得之于形外"和"计白当黑"之说。编排设计中留出适当的空白，能起到强调和引起注意，烘托主题，营造高雅的意境的作用。这种以虚衬实、虚实相生的关系成为编排设计必然的统一体。也有以文字为主的类似样宣的设计，则更加强调虚实以及疏密关系的运用，有的地方紧密，有的地方疏松，可谓"疏可走马，密不透风"，版面编排均衡而不呆板，张弛有致。当然，留白率较高的版面，适合于表达具高雅格调的资讯信息，稳健或严肃的机构形象；留白率较小的版面，适合于表达热闹而活泼，充满生机与活力的资讯信息。留白量的多少，可根据所表现的具体内容和空间环境而定。

THE GREATEST BOXER SHOWING OFF HIS RIGHT HOOK SHOT ON A LEICA

A YOUNG REBEL LEADER IN CUBA SHOT ON A LEICA

A NAVY SAILOR KISSING A STRANGER IN TIMES SQUARE SHOT ON A LEICA

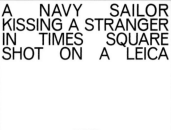

SOFORT INSTANT CAMERA. THE LEICA MADE FOR FUN.
LEICA STORE BOSTON   74 ARLINGTON ST

SOFORT INSTANT CAMERA. THE LEICA MADE FOR FUN.
LEICA STORE BOSTON   74 ARLINGTON ST

SOFORT INSTANT CAMERA. THE LEICA MADE FOR FUN.
LEICA STORE BOSTON   74 ARLINGTON ST

留白对于突出主体、创造意境有重要作用。
中国画中向来重视空白的运用，传统美学有
"形得之于形外"和"计白当黑"之说。

Project Modified Unit. 3 Journal is the result of 2 years of research outcomes conducted from MA Industrial design in Central Saint Martins. This journal contains the whole project journey including trial and error, experiments, methodologies learnt through the various workshops with organised and summarised format. Printed in the United Kingdom at Central Saint Martins, on Neenah Environment Birch 90gsm and G.F Smith C/P 100gsm.

002.   003.

系列文字海报设计
设计：Jack Forrest

Journal
Self-Directed design Research

Unit 3.

版面空间的下部，会给人以沉重、稳定的感受，设计时往往将各种具名性的信息，如企业名称、创建年月等信息放在下方。

　　编排设计的形式法则是创造美感平面空间的主要手法，起着活跃和统一版面、突出重点、贯穿前后等重要的视觉引导作用。每种形式原则，在表现上皆有不同的特点和作用，但在实际应用上都是相互关联而共同为用的。通过编排设计，设计者能让图文产生一种视觉引导，引导观者有序地观察第一视觉、第二视觉、第三视觉等，使内容主次分明，清晰有条理且具有一定的逻辑性，促进视觉信息快速、准确地表达和传播。对某一个特定信息界面的编排而言，各种元素的统筹和富有创意的表现，不仅是方便阅读的需要，也是产生视觉美感的需要，因此形式美的法则是影响版面编排优劣的决定性因素。设计师要根据主题构思一个类似音乐的"主旋律"，打破传统的约定俗成的做法，发挥创意和想象力，力求创造一个完美的"有意味的形式"。

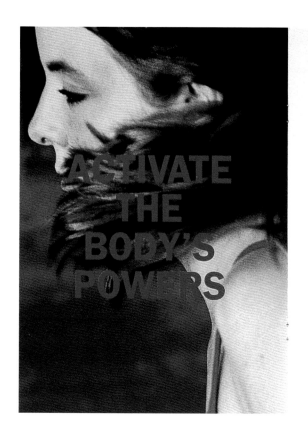

ACTIVATE THE BODY'S POWERS

SYMBOL OF LIFE

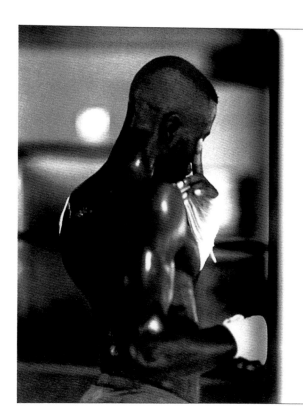

Knock Out! Photographs by The Douglas Brothers

七、编排设计逻辑

编排设计的一个重要的工作前提就是要对相关视觉流程进行分析，保证相关逻辑体系和编排信息的有效整合，使其符合人们的思维逻辑和视觉逻辑。创作者在收集视觉信息的过程中，要对视觉符号和数据基本结构进行整合，对感知信息进行接收和处理，借助相应的视觉逻辑结构有效整合控制机制，并对视觉逻辑的有序性进行多元处理，提升视觉逻辑的秩序化效果，以确保其符合实际标准。

1. 编排逻辑的整体秩序

在编排设计中，人们要结合美学要求，对编排设计的最终诉求进行快速整理，积极构建逻辑结构，从而保证设计的认知和理解效果更加有效，确保实际阅读感受更加有效。即在编排中，要对秩序性、韵律性以及受众的阅读习惯进行整合，在确保目的明确的同时，使相关视觉流程和编排效果达到最优化。

人们在建立视觉逻辑时，基本逻辑结构是从上到下、从左到右的，这就需要编排者聚焦核心内容进行统筹分析，保证受众接收信息的有效性。首先，要提高对视觉元素的注意程度，确保受众能有效地接收外界信息，可以第一时间对接收的信息进行判定，促进整个视觉观测效果的完整性。其次，要对具体问题进行集中整合，对有效性和直观性进行测试，对视觉逻辑的阅读性进行分析。这样才能从根本上提高设计元素的实际价值，为编排展开全面且有序的处理做准备。在实际编排设计中，要使信息能够有效且迅速地落实并进行传递，设计者需要对编排中不同的视觉元素进行分类管理，促进编排的合理有序，实现管理效果的全面优化，保证其视觉体系的完整度，达到信息层的完整性优化。

创意海报设计
设计：Yudai Osawa

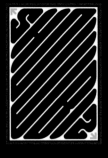

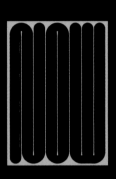

海报设计
设计：Berkay Taş
以英文字母作为主要的图形与编排元素。

2. 编排逻辑的有序层次

在对组织结构进行统筹，整合编排时，视觉的层次会影响信息的主次接收，设计者需要明晰信息要素的主谓宾，确保编排架构中的视觉符号更加符合有序的层次。

首先，将形的元素化解为最小单位的点、线、面，保证整体设计理念的清晰性。视觉是一种无形的关系，是设计结构和元素参数之间的有效关系。其次，对视觉流程进行处理，确保视觉效果和聚焦效果符合预期。再者，对编排过程和视觉感官效果进行整合，确保视觉流程和视觉元素之间能形成良好的互动，结合终极目标进行综合性处理，保证元素的优化和视觉流程的顺畅。在编排中，设计者也要借助形、色、质地等不同元素进行编排与组合，有效地建构完整的视觉层次，完成整体的编排设计，实现编排效果和具体因素的统筹控制。

此编排设计以鲜明、独特的形式美，清晰、准确地传达出丰富多彩的内容，创造出形神兼备的、具有生命力的视觉传达。

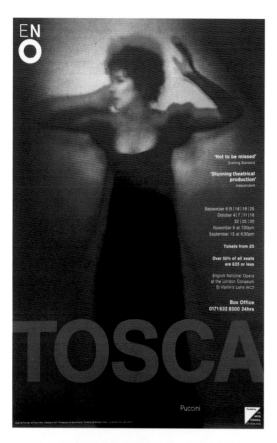

3. 编排逻辑的整合优化

　　编排需要建构差异化视觉，对编排的流畅性、优美性以及新颖程度予以重视，确保能清晰且直观地表达出相关情感或意境。编排以传达的主体思想为依据，将各种视觉要素进行科学的安排、组织和平衡协调。编排自始至终要抓住人们的视线，以"瞬时注目"为目的，比例恰当、主次分明，使视觉焦点处于最佳的视域，使观者能在瞬间感受主体形象的视觉穿透力。设计者要针对形态、色彩以及质地，对相关元素进行整合优化，运用编排有序合成，确保整个设计内容能发挥其理性价值，促进画面视觉逻辑的有效整合，实现编排和视觉之间良好的逻辑关系的形成。

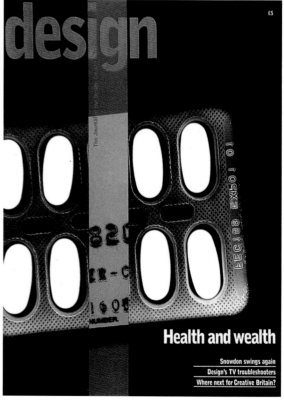

八、编排设计概览

华思正楷

设计：华思设计（刘永清、刘一墨）

此为端庄方正、字之楷秀的华思正楷的字体设计。版面的编排简洁，凸显文字的圆润清秀、结体严整、笔力挺拔、体势劲媚，笔画粗细均匀有力。

本色唱片设计
人物和唱片的色彩重复式排列，并且采用荧光色与黑色对比，凸显本色的主题。

A1) GET DOWN AND HOWL
A2) BENEATH THE SKIN
A3) I'M BORED
B1) BENEATH THE SKIN
(DANCEFLOOR DUB)
B2) MOVE FORWARD

NEUGRAU - BENEATH THE SKIN

Composed, recorded
and mixed by Ingmar Pauli and
Saša Rajković.

Mastered by Saša Rajković at
Sensorium studio. Design by
Michelangelo Greco.

SLK011
She Lost Kontrol Records
2020

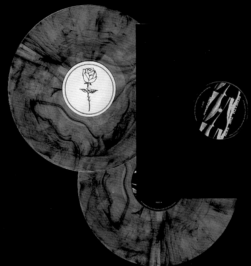

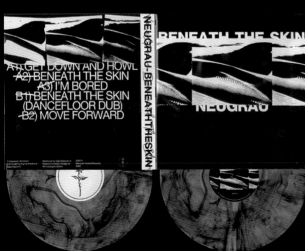

皮肤之下
设计：Michelangelo Greco

Andre Agassi

HAVE
MERCY.
KILL
QUICKLY.

Zoom
Air

is speed
cushioning

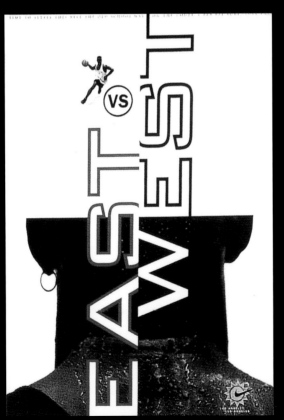

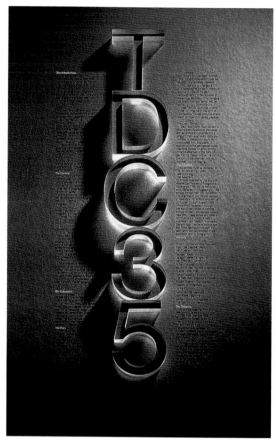

图形的多样化的组合、多层次的复叠、空间的延展、画面的肌理等使编排设计语言更为奇特绚丽。

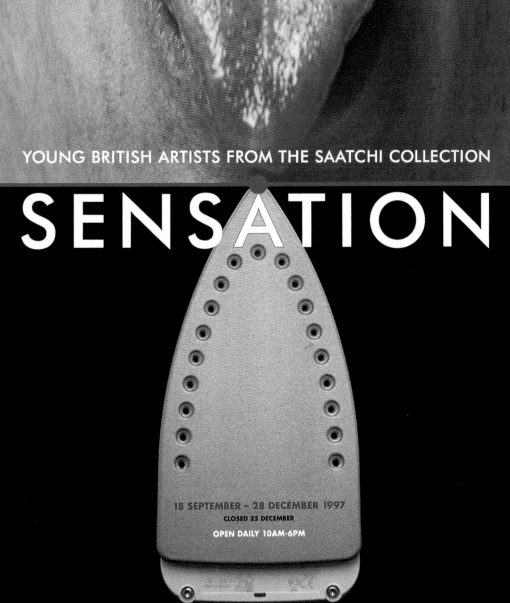

ROYAL ACADEMY OF ARTS
PICCADILLY LONDON W1

YOUNG BRITISH ARTISTS FROM THE SAATCHI COLLECTION

SENSATION

18 SEPTEMBER – 28 DECEMBER 1997
CLOSED 25 DECEMBER
OPEN DAILY 10AM-6PM

73

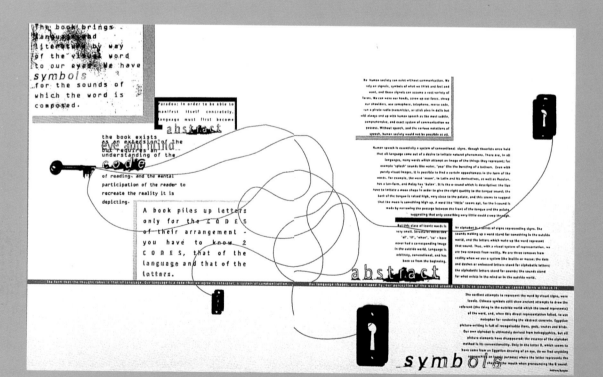

Systems not schemes, character
not conformity: typography, for
Erik Spiekermann, is a tool for
rendering the world accessible

Meta's tectonic man

Text: William Owen

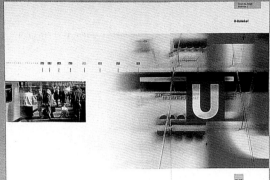

Erik Spiekermann is a consummate pluralist. Able to move, seemingly without effort, between roles as a typographer, designer, writer, public speaker and merchandiser, he was once even a politician – a Green Party member of the Berlin Senate. Spiekermann is the author of *Stop Stealing Sheep* and *Rhyme & Reason* – two models of typographic rectitude for a lay audience – and the articulate upholder of standards of public design in many a conference lecture. He is the designer of Meta, one of the most successful typefaces of this decade, and founder of the typeface distribution company FontShop. Spiekermann is also a partner in MetaDesign, now an international consultancy with offices in Berlin, London and San Francisco, and it is this manifestation – the graphic designer – that has received the least attention. For while Spiekermann has been busy promoting his own particular brand

"The page is the lowest common denominator of the book system. The page is the molecule and the atom is the word."

of rational Modernism-with-character, the polemicist has sometimes overshadowed the practitioner.

Spiekermann himself rejects the title of graphic designer to describe his practice: "I am a typographic designer. A typographic designer starts from the word up; a graphic designer starts from the picture down." This idiosyncratic explanation – few would place the two disciplines in opposition, and one is usually regarded as inclusive of the other – has a certain logic. But how, then, does Spiekermann distinguish his approach from that of an avowed graphic designer such as Gert Dumbar?

"Dumbar always uses space. He can't have three-dimensional space because paper is flat, so instead he uses cross-sections – he dissects objects in space and puts them on the flat page. He is a spatial kind of image guy; he thinks in theatrical terms. I think in page terms. The page is the lowest common

denominator of the book system. The page is the molecule and the atom is the word. You see, I read. I read before I design, and I write. I design outwards from words."

A respect for words and evident talent for using them might seem incongruous in the visual arts were it not that Spiekermann so often speaks and writes in pictures. His conversation is a stream of aphorism and metaphor. On national stereotypes in graphic design, for instance, we learn: "France is olive shaped; Holland is triangular, always very pointy and narrow; Germany is very square; and England is round." And on being a designer: "I am a servant, I'm not an artist. If I was an artist I would be oval, like an olive."

As a typographic designer, however, Spiekermann is distinctly quadrilateral. His trademarks are a rectangular or braced bar that bleeds off the page

and a palette of just two colours – black and red, in the craft tradition. While he is generous with words, Spiekermann is extremely parsimonious when dispensing colour, shape and typographic variation.

He ascribes this frugality to his background. The eldest of four children, his father a lorry driver, he was born into an impoverished post-war Germany, near Hannover, in 1947. He paid his way through Berlin's Free University, where he studied art history, by setting up as a jobbing letterpress printer. He had only two typefaces, in three sizes, and could rarely use more than two colours (he learned to give the impression of three-colour work by reversing type out of a colour in white). Economy was a function of necessity, and Spiekermann has used this experience to establish a method which exploits scarce resources to the full rather than attenuates or reduces. He is used to getting a lot from a little, but

1-3 Cover and pages from the design manual for the BVG, Berlin's city transport system. This was MetaDesign's first major corporate identity and public signage project, begun in 1987 with new livery, signage and timetables for the bus service, and enlarged to include the city railway system after unification in 1989. 4 BVG logotype. Uniquely, for MetaDesign, the device is symmetrical, centred within its background and based on an unsophisticated square.

The tenth pioneer

Seventeen and *Charm* both broke new ground. Cipe Pineles, their art director, played a leading part. Why, when the history came to be written, was Pineles left out?

1. Cipe Pineles and her husband William Golden, art director at CBS, receiving medals from the Art Directors Club of New York in 1948 – the first time a wife and husband were honoured in the same year. Pineles was awarded here for commissioning fine artists to illustrate Seventeen. 2. Charm's editors and executives celebrating its tenth anniversary in 1955. Seated from left, Eleanore Hildebrand Bruce, fashion editor; Helen Valentine, editor-in-chief; Estelle Ellis, promotion director; Cipe Pineles, art director. 3 (opposite). Pineles meeting at Charm, about 1955.

Text: Martha Scotford

Traditional accounts of design history have some obvious blindspots when it comes to the careers of women designers. Written from a predominantly male perspective, they tend to ignore the interactions of the personal and the professional, the private and the public, which play such a decisive role in the shaping of women's working lives. They are equally indifferent to the variety of career paths taken by women, the nature of collaboration between colleagues and the social and political roles played by women professionals.

The career of the American designer Cipe Pineles presents us with a much needed opportunity to explore new ways of writing design history. As an illustrator, design teacher and art director working primarily in women's magazines, she was an exemplary professional, an intriguing individual, and a valuable role model. Yet despite her many achievements, Pineles has never entirely received her due. The most recent historical study of individual American designers, R. Roger Remington and

Barbara J. Hodik's *Nine Pioneers in American Graphic Design* (1989), profiled nine male designers who worked in and around New York between the late 1920s and early 1970s and who "have helped shape graphic design as a profession and have made a distinctive and innovative contribution". Pineles' career fits the criteria for inclusion, yet though both her husbands – William Golden and Will Burtin – are among the chosen nine, she has been overlooked. Should Pineles have been the tenth pioneer?

Born to Jewish parents in Poland in 1908, Pineles came to the US in 1923 at the age of 15. Three years later she enrolled as a commercial art student at the Pratt Institute, followed by a year of painting supported by a Tiffany Foundation scholarship during which time she looked for design work. In a biographical summary note to Dr Robert Leslie, director of the Composing Room, in 1970, she described the frequent professional rejections she suffered when it was discovered she was a woman. Employers were reluctant to put her in the artists'

"Cipe Pineles was more than a token first woman art director; she was an innovator in art direction. She created the role of the modern woman designer and had no female peers"

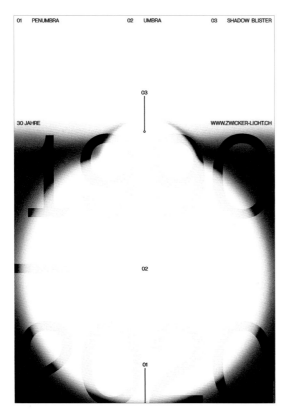

2020 德语区 100 张最佳海报

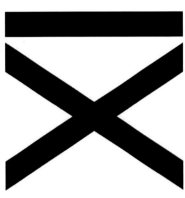

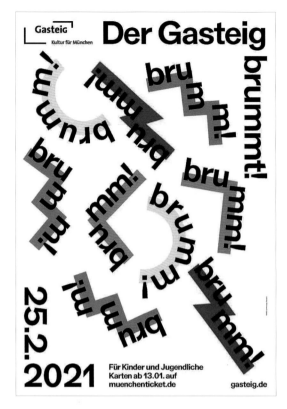

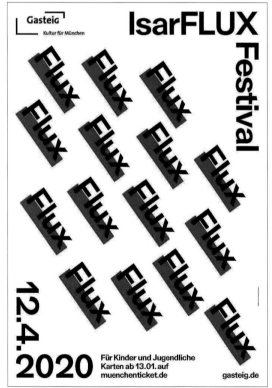

2020 德语区 100 张最佳海报

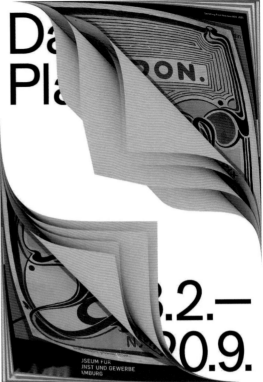

禁止砍伐，保护动物
设计：赵超
手写的满版的书法英文体现出呼吁与倡导。用树桩木纹形成各种
动物并结合书写的英文大标题来直观地表达，时刻提醒人们禁止
砍伐，保护动物，保护家园。

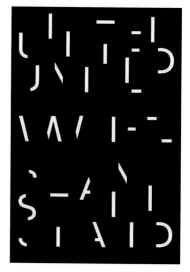

[编排设计]

纽约抗击新冠病毒海报设计

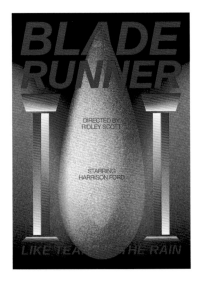

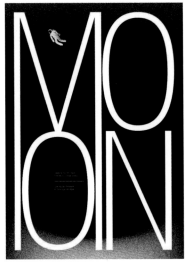

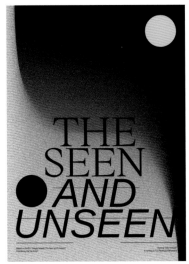

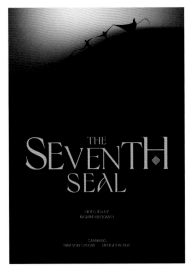

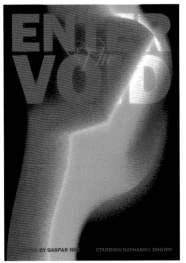

惊艳的色彩和图形构成设计

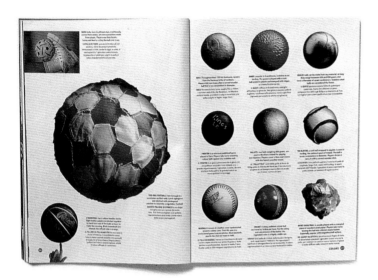

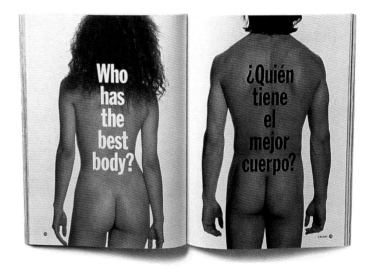

[编排设计]

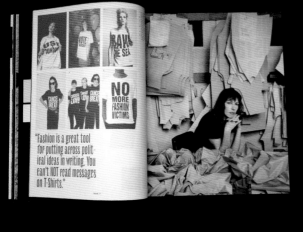

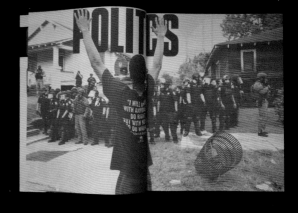

"Fashion is a great tool for putting across polit-ical ideas in writing. You can't NOT read messages on T-shirts."

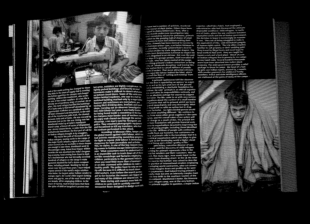

1. GLOBALISATION REBOOT 2. ASIAN TRAILBLAZERS
3. PREDICTABLY PREDICTABLE
4. GETTING PERSONAL

5. PLATFORMS FIRST 6. MOBILE OBSESSED
7. EVERYTHING GETS REAL
8. SUSTAIN CREDIBILITY

OUTSIDERS

Photographs by Robin Lewis Grierson

CLOSE UP

Anthony Hopkins as the Nazi Knight photographer Ivan Rassolt for a role in the Asia Playas film Amistad

Portraits by Gino Sprio

BIG

A~Z

Photographs by Marcus Tomlinson

OUISE DAHL-WOLFE

Courtesy Staley Wise Gallery, NY

POSTERS

BERLINER ENSEMBLE 62 — 99

26

Die Akademie der Künste der DDR zeigt

Filme aus Chile

17 Uhr
Der Schakal von Nahueltoro
20 Uhr
Liebe Genossen

Sonnabend 25. Juni 1983

Die Akademie der Künste der DDR präsentiert
Abenteuer Humor Phantasie
Trickfilme
Todor Dinow Karel Zeman

Kino für Knirpse mit Grips und für Große,
das Gruseln und Lachen nicht verlernt haben

So. 9.11. 15.00 So. 23.11. 15.00 Sa. 29.11. 15.00

Foyerausstellung Todor Dinow Karikaturen

Interview mit
Cesarina Drescher

Die Akademie der Künste der DDR zeigt
Preisgekrönte
KURZFILME
der Internationalen Festivals 1987
"Gestern ausgezeichnet!
Havanna Hiroshima Kraków
Oberhausen Moskau
LEIPZIG*
*Heute in Berlin!
Sa. 28. November 1987

15.00 17.30 20.00

Jeweils andere Filme!

15. September · 15.30 Uhr
Deutsche Staatsoper Berlin

Willens-
kundgebung

der Kultur- und Geistes-
schaffenden
zum 20. Jahrestag der
Gründung der
Deutschen Demokratischen
Republik

Brecht
Bücher
der DDR
Aus Anlaß des achtzigsten
Geburtstags von Bertolt Brecht
Ausstellung
9.Februar bis 3.März 1978

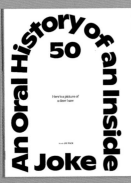

An Oral History of an Inside Joke

50

Here's a picture of a deer I saw

Words: JW PACK

"Humor can be dissected, as a frog can, but the thing dies in the process and the innards are discouraging to any but the pure scientific mind."
—E.B. White, 1941

Inside jokes are tricky. To those on, well, the inside, they're a secret language, a shared history, a time stamp with a punchline. The best ones are alive—they grow and mutate and get weirder and stupider and funnier as the years pass. They become the amateur comedian's best friend, a reliable bit. Who among us hasn't stood in front of a wedding, clammily clutching a microphone in one hand and a creased piece of paper in the other, recounting a story you know will get polite chuckles from 94 of the people there and tears of pure laughter from the other six? But for anyone on the outside of an inside joke? Uggggggh. On rare occasions, however, and against all odds, an inside joke can transcend its own exclusivity; it can get *everyone* howling like those six idiots. Lucky for you, "Here's a Picture of a Deer I Saw" is one of those. To get the joke, you need the story, and the story of "Here's a Picture of a Deer I Saw" is a bonkers one. It's a matryoshka doll of shenanigans: an inside joke nestled inside another inside joke which is itself wrapped in yet another. And it's not even my story, or my joke. It belongs to my friend Jenna and her friends, and I've just co-opted the shit out of it. I've told it at dinner parties and to my family and to my coworkers. I'm about to tell it to you, with help from everyone else who loves it as much as I do. Ready? Here's a picture of a deer I saw!

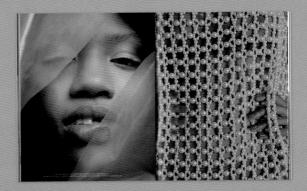

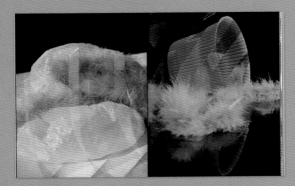

The Archive

of Future Land-scapes

HOW TO BURY A TREASURE

8

Words: CONA SOLER

Stir Crazy

Nothing is sacred

Photography: PLUM McCLEAN

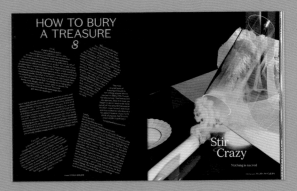

Gossamer 杂志内页排版设计

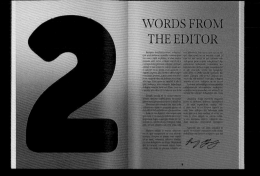

WORDS FROM THE EDITOR

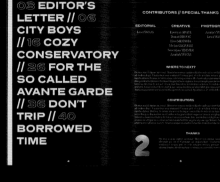

CONTRIBUTORS // SPECIAL THANKS

EDITORIAL   CREATIVE   PHOTOGRAPHY

WHERE TO NEXT?

CONTRIBUTORS

THANKS

2

THE CITY BOYS

REBELS WITHOUT A CAUSE

IF YOU REALLY WANT TO ESCAPE THE THINGS THAT HARASS YOU, WHAT YOU'RE NEEDING IS NOT TO BE IN A DIFFERENT PLACE BUT TO BE A DIFFERENT PERSON.

IF THE MUSE EXISTS, SHE DOES NOT WHISPER TO THE UNTALENTED.

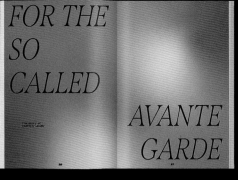

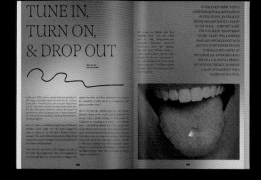

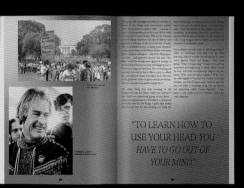

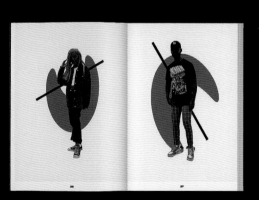

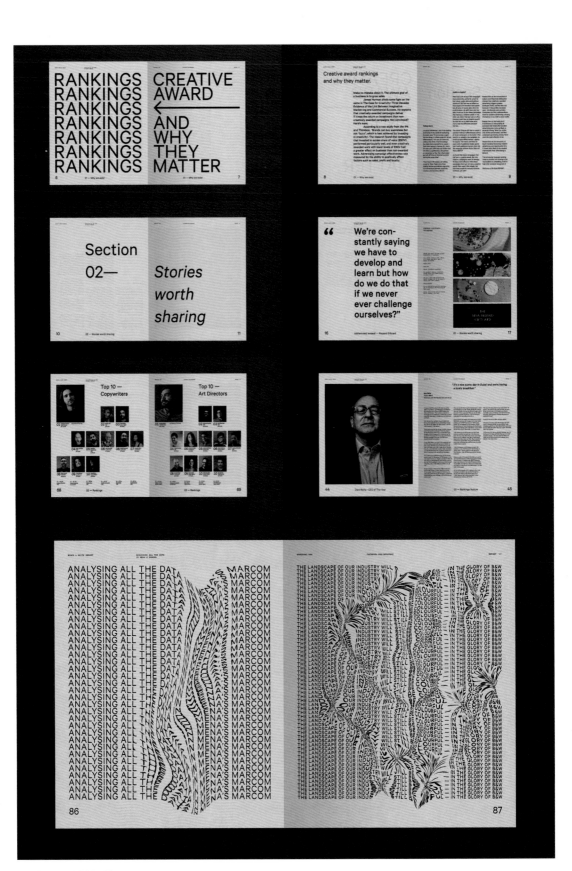

The Black And White Report
设计：Ryan Atkinson

Journal

Self Directed design Research

Unit 3.

Project Modified Unit. 3 Journal is the result of 2 years of research outcomes conducted during MA Industrial design in Central Saint Martins. This journal contains the whole project journey including trial and error, experiments, methodologies learnt through the various workshops with organised and summarised format. Printed in the United Kingdom at Central Saint Martins, on Neenah Environment Birch 90gsm and G.F Smith C/P 100gsm .

002.   003.

Index

Intro-duction

006.   007.

(1-B)   Key Area of Investigation

The project demonstrates the possibilities of making new daily products by using the available parts of obsolete smartphones and guides individuals to participate in this. Plus, by providing instruction which guides users to follow to make products, it gets closer to the project aim, which is "increasing a product's value from individuals' fingertip".

The structure of this Journal shows the project progression. In Stage 1, through contextual review, the situation surrounding the topic is explored to discover the direction and possibility of the project. Next, a standpoint and a research question are set, shaping the project. Through Stage 3, shows the testing, experimentation and exploration by applying various design methodologies to reach the project outcomes. Stage 4 deals with the result. Finally, Stage 5 illustrates critical reflection upon the two years MA journey.

(1-C)   Project Overview + Report Structure

The project demonstrates the possibilities of making new daily products by using the available parts of obsolete smartphones and guides individuals to participate in this. Plus, by providing instruction which guides users to follow to make products, it gets closer to the project aim, which is "increasing a product's value from individuals' fingertip."

The structure of this Journal shows the project progression. In Stage 1, through contextual review, the situation surrounding the topic is explored to discover the direction and possibility of the project. Next, a standpoint and a research question are set, shaping the project. Through Stage 3, show the testing, experimentation and exploration by applying various design methodologies to reach the project outcomes. Stage 4 deals with the result. Finally, Stage 5 illustrates critical reflection upon the two years MA journey.

010.

Critical Raw Materials

(1)   Beryllium, found in Connectors
        and Springs but being phased out,
        Recycling rate 0%
(2)   Cobalt, found in batteries,
        Recycling rate 35%
(3)   Indium, found in touch screen,
        recycling rate 0%
(4)   Magnesium, found in metal alloy in
        casing, recycling rate 0%
(5)   Silicon Metal, found in microchips,
        recycling rate 0%
(6)   Tantalum, found in
        microcapacitors, recycling rate 1%
(7)   Gallium, found in Integrated Circuits,
        recycling rate 0%
(8)   Graphite, found in Batteries,
        recycling rate 0%
(9)   Neodymium, found in microphones,
        recycling rate 0%
(10) Rare earth elements, found in
        camera, recycling rate 0%
(11) Tungsten, found in vibratone
        motors, Recycling rate 42%

086.   087.

(2-C)   E-WasteEffort Against E-Waste

Recently, the whole world has paid attention to this problem, and there are some efforts to overcome it. One of them is Urban mining, which is a recycling system for producing precious metals from e-waste in an urban area (Gruenewelle, No date). For the second example, Restart Project is a social enterprise based in London, which aims to provide a place where people teach how to fix broken and slow electronic devices each other (Restart, 2019).

→ P. 87   Lastly, Apple also has launched Daisy, which can disassemble 15 different iPhone models at a rate of 200 per hour, recovering even more important materials for reuse (Apple Press, 2019). For the further research direction, I had a question which is: how can an individual who creates E-Waste directly contribute to this problem? And I wanted to focus on consumption behaviours related to mobile phone e-waste.

020.   021.

(1)   Urban Mining
        An urban mine is the stockpile of rare
        metals in the discarded waste electronic
        and electronic equipment (WEEE) of a
        society.[1] Urban mining is the process
        of recovering these rare metals through
        mechanical and chemical treatments.
(2)   Restart Project (2018)
        The Restart Project was born in 2013
        out of our frustration with the throwaway,
        consumerist model of electronics that
        we've been sold, and the growing mountain
        of e-waste that it's leaving behind. By
        bringing people together to share skills and
        gain the confidence to open up their stuff,
        it gives people a hands-on way or making a
        difference, as well as a way to talk about the
        wider issue of what kind of products

(3)   Apple's Daisy' Recycling
        Robot (2019)
        Daisy, Apple's latest innovation in
        material recovery, can disassemble nine
        different iPhone models to recover valuable
        materials that traditional recyclers cannot.

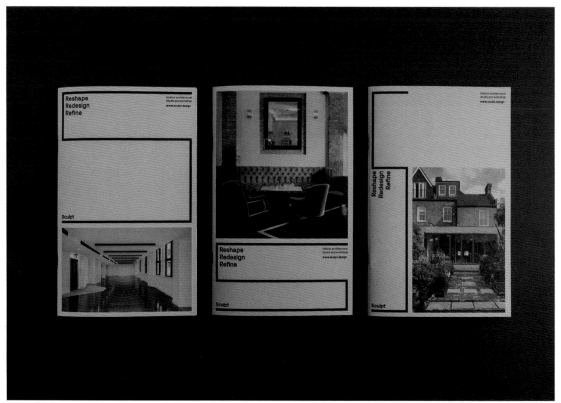

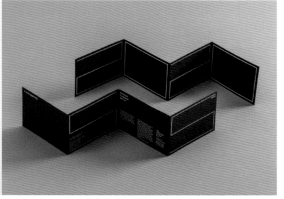

建筑设计工作室 Sculpt 品牌视觉设计
设计：Common Curiosity

HUNTERIAN ART GALLERY, UNIVERSITY OF GLASGOW

William Whitfield & Partners

'BRUTALIST ARCHITECTURE WAS MODERNISM'S ANGRY UNDERSIDE'

St Andrews RC Church, Livingston

Baxter Clark and Paul

LOCATED ON GLEN ISLA
56°43'47.3"N 3°13'18.4"W

MATTHEW BUILDING, UNIVERSITY OF DUNDEE

Baxter Clark and Paul

BÉTON BRUT
BÉTON BRUT
BÉTON BRUT
BÉTON BRUT
BÉTON BRUT
BÉTON BRUT
BÉTON BRUT
BÉTON BRUT
BÉTON BRUT
BÉTON BRUT

TURB NOT̃ÈB

56°27'23.3"N 2°59'01.0"W

GENERAL ACCIDENT WORLD HQ PERTH

James Parr & Partners,

TOTAL COST
30,000,000

MCIVER HOUSE

R. Seifert and Partners,

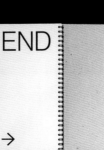

END

→

Barcelona Novel·la Històrica

Petites Històries que han fet història

4-9 nov. 2019

Isabel Allende,
José Enrique Ruiz–Domènec,
Jo Baker,
Rafel Nadal,
Carme Riera,
Isabel García Trócoli,
Care Santos,
Anelio Rodríguez…

Ajuntament de Barcelona   BARCELONA CULTURA   Barcelona Ciutat de la Literatura UNESCO

#BCNhistorica19
barcelona.cat/novel.lahistorica
bcnliteratura   @bcnliteratura

Barcelona Novel·la Històrica

Petites Històries que han fet història

4-9 nov. 2019

2019
Setena edició

Índex

Els historiadors s'interessen cada vegada més per la petita història, la història que escriu, sense saber-ho, la gent del carrer, homes i dones allunyats de l'esfera pública que participen de manera activa en fets que marquen el progrés d'una societat. I cada vegada més, també, aquesta petita història conserva entre les cartes i records i documents dels seus grans, conscient del seu enorme valor humà i històric. Perquè entre les cartes, les fotografies i els dietaris de les generacions que ens han precedit podem trobar informació de primera mà sobre costums i tradicions, sobre rutes ignorades pels manuals d'història o transformacions urbanes.

Gràcies a la generositat de persones que han cedit el seu arxiu familiar per a la consulta i l'estudi, en els darrers temps, els arxius públics de Barcelona s'han anat engrossint. Fel·lícules domèstiques, documents oficials, postals, expedients o retalls de diaris han restaurat i catalogat perquè tothom pugui consultar-los. I no només els historiadors, sinó també els novel·listes, s'han nodrit d'aquestes fonts, autèntiques i legítimes, per recuperar episodis oblidats de la nostra història i personatges silenciosos.

La gran literatura inspirada en les petites històries quotidianes centra aquesta edició de Barcelona Novel·la Històrica. Una literatura que en tribut a les persones que han pres part en successos rellevants però no han estat objecte d'homenatges ni memòries. Una literatura que restitueix la gent corrent, anònima, en el lloc que li correspon en la història. La ciutadania, el poble, n'és el protagonista.

Joan Subirats
Tinent d'alcaldia de Cultura, Educació, Ciència i Comunitat

3

Dilluns 4
18.00 h

Ajuntament de Barcelona
Saló de Cent
Plaça de Sant Jaume, 1

Inaugura:
Isabel Allende

Lliurament del Premi Internacional de Novel·la Històrica Barcino

Barcelona Novel·la Històrica inaugura la seva setena edició al Saló de Cent de l'Ajuntament de Barcelona, escenari, com sempre, de l'acte de lliurament del Premi Internacional de Novel·la Històrica Barcino. El guardó, que reconeix el valor literari d'un autor o una autora que ha conreat el gènere, celebra enguany la literatura d'Isabel Allende, creadora d'històries inspirades en destacats fets històrics i una de les escriptores més llegides del món.

Els successors Enric Calpena i Care Santos, els comissaris Sergi Doria i Óscar López, Fèlix Riera, comissari de Barcelona Novel·la Històrica, componen el jurat de l'edició. Anteriorment, havien estat premis Barcino Lindsey Davis, Santiago Posteguillo, Simon Scarrow, Christian Jacq, Arturo Pérez-Reverte i Leonardo Padura.

Dilluns 4
19.15 h

Ajuntament de Barcelona
Saló de Cent
Plaça de Sant Jaume, 1

Care Santos, autora de Medik vita
i Gènere blau, entre altres novel·les
històriques i José Enrique Ruiz-
Domènec, historiador, autor d'Entre
historias de la Edad Mèdia i Ricard
Guillem o el somni de Barcelona.
Moderat:
Fèlix Riera, comissari
de Barcelona Novel·la Històrica

Històries

Les històries anònimes, en ser revelades, ens permeten entendre millor la història i ens fan veure que, potser, no és ben tal com ens l'havien explicat. L'obra d'historiadors com Georges Duby, Carlo Ginzburg o Natalie Zemon Davis, si la d'escriptors com Umberto Eco, exemple que l'estudi dels esdeveniments històrics ens porta a descobrir en els detalls, en les anècdotes, en les vides d'artesans, camperols i comerciants o en les llegendes populars, una història de gestes particulars, silencioses i sovint inverídes. Una història que completa i enriqueix l'oficial i, de vegades, fins i tot l'esmena.

8   9

Divendres 8
19.00 h

Biblioteca Jaume Fuster
Plaça de Lesseps, 20-22

Autors participants: Víctor Amela,
Montse Santos, Andreu Claret,
Oriol Comas i Coma, Berta Jané,
Xavier Moret, Rafel Nadal, Adrià
Pujol, Gerard Quintana, Pilar
Romero i Joan Santanach.

Cronos, acte festiu amb lectors i escriptors

A mig camí entre un acte literari i un de lúdic, Cronos posa el punt final a una nova edició de Barcelona Novel·la Històrica. Màrius Serra, filòleg i lactobiòsula, ens convida cada tardor a participar al Cronòstic, un joc d'endevinament que posa a prova els coneixements històrics dels assistents.

La mecànica és senzilla: selecció d'autors i autores llegeix fragments breus de les seves novel·les històriques i, tot seguit, el públic la una travessa per pronunciar l'any en què ocorre cada fragment. L'obsectiu final és endevinar la mitjana exacta entre tots els anys del Cronograma.

Rutes literàries amb Martí Gironell

Dimarts, 5, a les 16.30 h
Dijous, 7, a les 16.30 h
Dissabte, 9, a les 10.00 h
i a les 12.00 h

Lloc de trobada:
Escoltor Huguet de la Universitat de Barcelona, Gran Via de les Corts Catalanes, 585

Cal inscripció prèvia a:
arhos.cat/bcn.cat
Màx. 4 dies entrades per sessió.
Cal reservar el dia que es vol assistir
a l'itinerari, el màxim i espera de
les persones apuntades.

Petita ruta per la gran història

El periodista i escriptor Martí Gironell proposa en aquesta edició de Barcelona Novel·la Històrica una ruta literària que permet recuperar petites històries que han ajudat a escriure la història.

La Universitat de Barcelona, a la Gran Via, és el punt de partida de la ruta i, també, l'escenari d'Els impostors, novel·la de Pilar Romera que ret tribut als professors d'universitat foragitats després de la guerra. Bon a prop, a l'Hotel Regina, Joaquim Folch i Torres, futur director del primer MNAC, s'entrevista amb Ignasi Pollack, antiquari i marxant d'art, per rescatar les obres del romànic català als Pirineus; com al 1918 i rescarva ens remet a la novel·la Stradoppu, del mateix Martí Gironell.

Jo hauria pogut salvar Lorca, de Víctor Amela, recorda el recital que Lorca i Margarida Xirgu van oferir al Teatre Barcelona, avui desaparegut. Una multitud va omplir el teatre i el carrer, rambla de Catalunya amb entra de Sant Pons. La ruta enllaça uns el carrer de Mirallers, al barri de Sant Pere, Santa Caterina i La Ribera, un Cristòfor Colom tenia casa, a L'enigma Colom, de Maria Carme Roca ens traslada a la Barcelona del 1493 i la seva conspiració que en travel·la per fer fracassar el gran projecte de l'almirall. L'última etapa de l'itinerari ens porta al fossar de Víctor Jurado Riba, empren un vaixell de jutge del 1714 per conèixer els homes i les dones anònims que van fer la història.

20   21

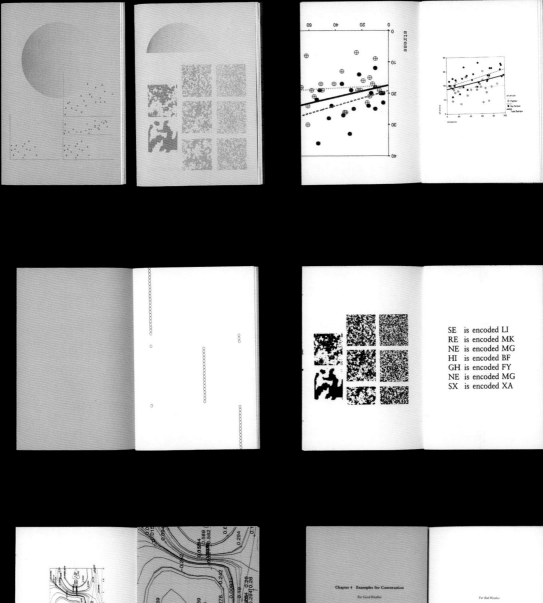

SE is encoded LI
RE is encoded MK
NE is encoded MG
HI is encoded BF
GH is encoded FY
NE is encoded MG
SX is encoded XA

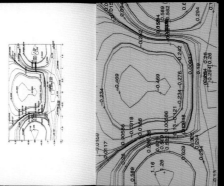

Chapter 4 Examples for Conversation

For Good Weather

"Nice day, isn't it?"
"Isn't it beautiful?"
"The sun..."
"Isn't it wonderful?"
"Yes, wonderful, isn't it?"
"It's so nice and hot..."
"I think it's so nice when it's hot, isn't it?"
"I really love it, don't you?"

For Bad Weather

"Terrible day, isn't it?"
"Isn't it unpleasant?"
"The rain ... I don't like the rain."
"Just think - a day like this in July. It rains in the morning, then a lot of sun and then rain, rain, rain, all day."
"I remember the same July day in 1936..."
"Yes, I remember too."
"Or was it 1928?"
"Yes, it was."
"Or in 1938?"
"Yes, that's right."

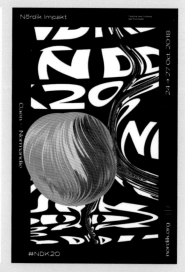

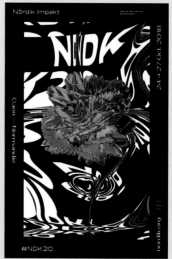

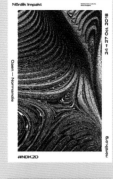
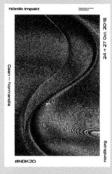

Nördik Impakt 20
设计：Murmure

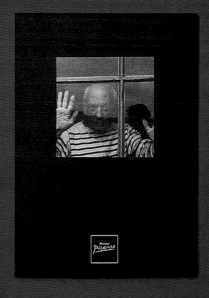

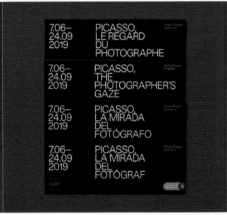
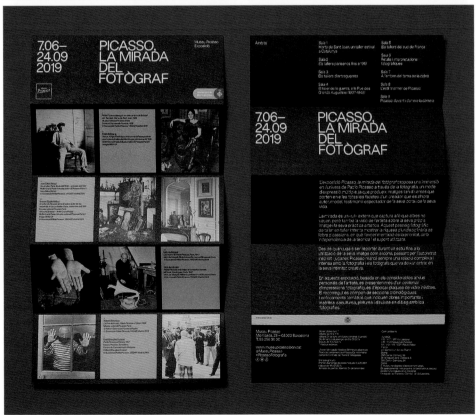

《毕加索，摄影师的凝视》
设计：PFP, Disseny

Trans y Cyborg, TALLER – AR Flye
设计：Saul & Co

北平机器
设计：窦誉笙

编

ART-
ZAVOD
PLAT
FORMA

КИЕВСКИЙ
ЛАЙФСТАЙЛ
НА ВЫХОДНЫЕ

АРТ-ЗАВОД
ПЛАТФОРМА — ЦЕНТР
РАЗВИТИЯ IT-ИНДУСТРИИ,
ОБРАЗОВАНИЯ, МУЗЫКИ,
ДИЗАЙНА, МОДЫ И
ИСКУССТВА. ЭТО КОНЦЕПТ
«ГОРОД В ГОРОДЕ»

ПЯТНИЦА
25 АПРЕЛЯ 2018

3.7 ■ ART SPACE

Открытие сезона Кураж
Базара пройдет 24 и 25 марта!
Снимай пуховик, обувай любимые
кроссовки, у нас Китайский Новый
год: весна, музыка и танцы все
выходные напролет.

Конкурс юных
альтернативных кофейщиков. Для
всех кто умирает за кофе, ведь это
такой же наркотик как и все
остальное в нашей жизни.

СУББОТА
26 АПРЕЛЯ 2018

2 ■ ART DOM
11:00—19:00

Лапы и котлеты — конкурс
собаководов и любителей питомцев.
Пробная и детальная инфа на сайте

4 ■ FOOD COURT
с утра до вечера

Уличная еда — уже лето, а
значит поздно думать про фигуру.
Ешь, пей, гуляй пока молодой. Не
думай не о чем просто кушай
криветки и мидии.

12 ■ COFFEE CENTER
с 20:00 — 23:00

Конкурс — юных кофейщиков,
которые любят делать альтернативные
напитки.

ВОСКРЕСЕНЬЕ
27 АПРЕЛЯ 2018

3 ■ ART DOM
11:00—19:00

Путешествия с Лунтиками —
детские развлечения целый день
нон-стоп.

4 ■ FOOD COURT
с утра до вечера

Уличная еда — уже лето,
а значит поздно думать про фигуру.
Ешь, пей, гуляй пока молодой.
Монатик, Онука, Касабиан, Воппи
Видоплясова Арка, Бйорк и другие
артисты украинской эстрады.

ART-
ZAVOD
PLAT
FORMA
AFISHA
2018

ART-ZAVOD
PLATFORMA
.COM

ДАЙДЖЕСТ
НА 25 — 27
АПРЕЛЯ 2018

КИЕВСКИЙ
ЛАЙФСТАЙЛ
НА ВЫХОДНЫЕ

АРТ-ЗАВОД
ПЛАТФОРМА — ЦЕНТР
РАЗВИТИЯ IT-ИНДУСТРИИ,
ОБРАЗОВАНИЯ, МУЗЫКИ,
ДИЗАЙНА, МОДЫ И
ИСКУССТВА. ЭТО КОНЦЕПТ
«ГОРОД В ГОРОДЕ»,
СОЗДАННЫЙ ДЛЯ РАБОТЫ
И ОТДЫХА.

Это наша миссия и мы готовы с вами
связаться по этому поводу

ВОСКРЕСЕНЬЕ
27 АПРЕЛЯ 2018

СУББОТА
26 АПРЕЛЯ 2018

2 ■ ART DOM
11:00—19:00

Лапы и котлеты — конкурс
собаководов и любителей питомцев.
Пробная и детальная инфа на сайте

4 ■ FOOD COURT
с утра до вечера

Уличная еда — уже лето, а
значит поздно думать про фигуру.
Ешь, пей, гуляй пока молодой. Не
думай не о чем просто кушай
криветки и мидии.

12 ■ COFFEE CENTER
с 20:00 — 23:00

Конкурс — юных кофейщиков,
которые любят делать альтернативные
напитки.

ВОС
27 АПРЕ

2 ■ ART DOM
11:00—19:00

Путеш
детские развле
нон-стоп.

С детьми буду
аниматоры, ко
о чаде и пойти
Посмотри ниж
на фуд корте.

4 ■ FOOD COU
с утра до вечер

Улична
а значит поздн
Ешь, пей, гуляй
Монатик, Онук
Видоплясова А
артисты украи

Все рестораны
сегодня будут

ART-
ZA
PLAT
FO

КИЕВСКИЙ
ЛАЙФСТАЙЛ
НА ВЫХОДНЫЕ

краткая гусиная
энциклопедия:
**гусь в мифологии ●
медумские гуси ●
гусь как сексуаль-
ный символ ● со-
бери гуся ● цифры
● сократ клянется
гусем ● гуси рим
спасли ● лучшая
в мире подтирка**

ГУСЬ В МИФОЛОГИИ

МЕДУМСКИЕ ГУСИ

ГУСЬ КАК СЕКСУАЛЬНЫЙ СИМВОЛ

ЦИФРЫ

- Во время линь-
ки птицы теряют
маховое опере-
ние. Из-за чего
1,5 месяца
не могут летать.

- Гуси использу-
ют около 10 зву-
ков для общения,
в зависимости
от ситуации.

- Число яиц
обыкновенно
6—12шт.

- Весят до 8 кг.

- Уже через 24
часа, вылуплен-
ные из яйца ма-
ленькие гусята
плавают наравне
с мамой.

- Средняя про-
должительность
жизни гусей
составляет
около 25 лет.

- У гусей доста-
точно хороший
слух, позволяю-
щий различать
звуки на рассто-
янии не менее
50 метров.

- Гусей одомаш-
нили в Древнем
Египте более
4000 лет назад.

СОКРАТ КЛЯНЕТСЯ ГУСЕМ

ГУСИ РИМ СПАСЛИ

ЛУЧШАЯ В МИРЕ ПОДТИРКА

ГАЙ РИЧИ

ДЕНЬ КИНО

18 МАРТА
УДРО 11 ОКТЯБРЬ
УЛ. НОВЫЙ АРБАТ 24

КАРТЫ, ДЕНЬГИ, ДВА СТВОЛА
БОЛЬШОЙ КУШ
РОК-Н-РОЛЬЩИК
МЕЧ КОРОЛЯ АРТУРА

$931 000 000
E931 000 000
$9 713 929

СБОРЫ В США

Джейсон Флеминг
Декстер Флетчер
Ник Морган
Джейсон Стэйтем
Стивен Макинтош
Николас Роу
Ник Марк и др.

На свой первый съёмочный день Винни Джонс отправился прямиком из полицейского участка. Он был арестован за избиение своего соседа.

Джейсон Стэйтем до знакомства с Гаем Ричи работал уличным продавцом. В самом начале картины ему пришлось сыграть самого себя в бывшей ипостаси.

Гай Ричи никак не мог определиться с концовкой своего фильма, поэтому её пришлось снимать несколько раз. Именно по этой причине голову Тома покрывает головной убор. За время, прошедшее с момента, казалось бы, окончательного завершения съёмочного процесса, актер успел отрастить волосы и наотрез отказался их укорачивать.

КАРТЫ, ДЕНЬГИ, ДВА СТВОЛА
БОЛЬШОЙ КУШ
РОК-Н-РОЛЬЩИК
МЕЧ КОРОЛЯ АРТУРА

2008
000 000
$728 089
$18
$25

Режиссер картины Гай Ричи появляется в камео, проезжая на велосипеде мимо Романа и Микки, когда они приближаются к дому Джонни.

В фильме присутствуют только 12-долларовые купюры, в реальности не существующие.

В фильме звучит одна из песен группы «Ех-Сектор Газа» — «Допился». Причем лидер группы Игорь Кущев узнал о том, что его композиция попала в фильм, из интернета.

Пулемётный монтаж в сексуальной сцене был сделан из-за того, что в день съёмок у Джерарда Батлера была инфекция дыхательных путей, и Тэнди Ньютон отказалась с ним целоваться. Гаю Ричи пришлось импровизировать.

Джерард Батлер
Том Уилкинсон
Тэнди Ньютон
Марк Стронг
Идрис Эльба
Том Харди
Карел Роден и др.

РОК-Н-РОЛЬЩИК
МЕЧ КОРОЛЯ АРТУРА

*pography Program BBE 2.0 — 3.0 编排设计

...авший Тайрона, пришел на съёмочную площадку фильма, ища качестве охранника. После того, как его увидел Гай Ричи, же предложил ему роль Тайрона.

...и слово «fuck» произносится 153 раза.

...чтобы поддерживать порядок на съёмочной площадке, Гай Ричи ...еобразную систему штрафов, среди которых были штрафы ...щие мобильные телефоны, опоздание, сон в рабочее время, ...зоры, за выражение недовольства...

2000
ГОД

$10 000 000
БЮДЖЕТ

$83 557 872
СБОРЫ В МИРЕ

《想象一下这个》书籍设计
设计：David Rindlisbacher

Slanted Magazine #36 — COEXIST
设计：Slanted Publishers

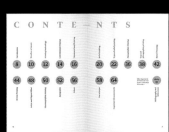

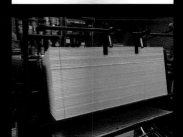

Beginner's Guide to Printing

设计： Anuvrat Dahiya

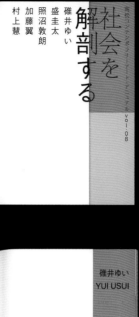

高松コンテンポラリーアートアニュアル vol.08

社会を解剖する

碓井ゆい
盛圭太
照沼敦朗
加藤翼
村上慧

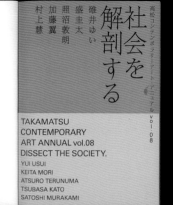

高松コンテンポラリーアートアニュアル vol.08

社会を解剖する

TAKAMATSU
CONTEMPORARY
ART ANNUAL vol.08
DISSECT THE SOCIETY.

YUI USUI
KEITA MORI
ATSURO TERUNUMA
TSUBASA KATO
SATOSHI MURAKAMI

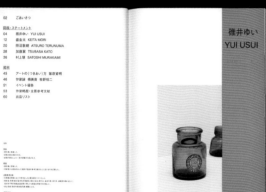

碓井ゆい
YUI USUI

碓井ゆい YUI USUI

I am exhibiting two works in this exhibition.

盛圭太
KEITA MORI

剖析社会的艺术书籍
设计：Keiji Yano

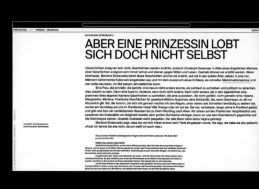

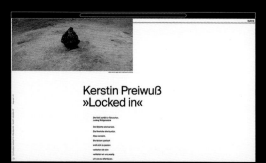

宣传册设计
设计：Daniel Zenker

01 TYPEFACE

Text example: Slab Serif, Slant, Semi Contrast

A *A* A
a *a* a
(a)@

A a *A a*

Text example: Weight / Slant

ABCDEFGHIJKLMNOPQRSTUVWXYZ *ABCDEFGHIJKLMNOPQRSTUVWXYZ*
abcdefghijklmnopqrsßtuvwxyz *abcdefghijklmnopqrsßtuvwxyz*
ABCDEFGHIJKLMNOPQRSßTUVWXYZ *ABCDEFGHIJKLMNOPQRSßTUVWXYZ*
abcdefghijklmnopqrsßtuvwxyz *abcdefghijklmnopqrsßtuvwxyz*
ABCDEFGHIJKLMNOPQRSßTUVWXYZ *ABCDEFGHIJKLMNOPQRSßTUVWXYZ*
abcdefghijklmnopqrsßtuvwxyz *abcdefghijklmnopqrsßtuvwxyz*
ABCDEFGHIJKLMNOPQRSßTUVWXYZ *ABCDEFGHIJKLMNOPQRSßTUVWXYZ*
abcdefghijklmnopqrsßtuvwxyz *abcdefghijklmnopqrsßtuvwxyz*

BAGAGE

ABCDEFGHIJKLMNOPQRSTUVWXYZ
aabcdefgghijklmnopqrrstuvwxyz
0123456789o123456789
""'¿'·'#/\()[]----«»··
ÁÀÂÄÃÅĀĂÆ·ÇĆČĊĎĐÈÉÊËĒĚĘ·ĜĞH
ĲĨĬĨĶĹĻŁÑŃŇŅÒÓÔÖÕŌŐØŒ·ÞŔŘR
ŚŠŞẞŢŤŦÙÚÛÜŪŮ·Ŵ·ẂỲÝŶŸŹŻŽ
áàâäãåāăæ·çćçčd·dèéêëēěęĝ ĝh·íìíiii
iĳjjkĺll·ñńňņ·òóôöõōőøþŕřŕśšşẞtťŧ·
ùúûüūůẃ·ẃỳýŷ·źżž·àóàáåâ ĝŕŕ·
ff fi fl ffi ffl
+−×÷=≠<>≤≥±¬~∑ %‰
¤□$€£¥†→↗↘←↙¢@&¶§©®™%[]

Text example: Sans Serif

ABCDEFGHIJKLMNOPQRSßTUVWXYZ
ABCDEFGHIJKLMNOPQRSßTUVWXYZ
abcdefghijklmnopqrsßtuvwxyz
ABCDEFGHIJKLMNOPQRSßTUVWXYZ
abcdefghijklmnopqrsßtuvwxyz
ABCDEFGHIJKLMNOPQRSßTUVWXYZ
abcdefghijklmnopqrsßtuvwxyz
ABCDEFGHIJKLMNOPQRSßTUVWXYZ
abcdefghijklmnopqrsßtuvwxyz
ABCDEFGHIJKLMNOPQRSßTUVWXYZ
abcdefghijklmnopqrsßtuvwxyz

5-axis Variable Font & Custom Font Generator

One font to rule them all

Donald E. Knuth,
Douglas R. Hofstadter,
Gerrit Noordzij,
Johanna Drucker,
Ellen Lupton,
Alan M. Turing,
Klasse Hickmann.

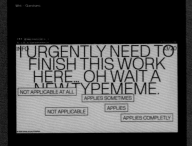

Web - Questions

Web - Questions

Web - Info

Landingpage Mobile

Landingpage Web

A	VORWORT		7
	THEORIE		11
B	SCHRIFT		27
	SYNOPSIS		47
C	ALGORITHMUS		269
	WEB-APPLICATION		277
	CODE		293

Speculative Type Design 推测式设计
设计：Daniel Stuhlpfarrer

The Caves of Steel

New York City → Overpopulated

Humanity is now shielded by a huge dome.

Once again living in caves, as the primitive humans did in the past.

Robots are taking over the work force.

Do they actually replace humans, or only help them advance developing?

Fritz Lang

ISAAC ASIMOV

Inspiring SciFi Creators

Hideo → Kojima

Ridley Scott

Philip K. Dick

Katsuhiro Otomo

Who said getting to the moon would be easy?

1988

Space Shuttle Challenger explosion

The Space Shuttle Challenger disaster is the biggest catastrophe in the history of spaceflight, which also changed the public perception of the US space program. All seven crew lost their lives, which included a civilian astronaut for the first time. The explosion shortly after take-off was broadcasted live on national television and left a mark in the public.

Challenger

Nerv

The Second Impact is a cataclysmic event which occurred in Antarctica on September 13, 2000. It was caused by an experiment conducted on Adam by a group of scientists led by Dr. Katsuragi.

Neon Genesis Evangelion

Eva Unit-01

Piloted by Shinji Ikari

Human Instrumentality Project

Spicy Italics

LET'S SEE WHAT KIND OF NIGHTMARES TOMORROW'S TECHNOLOGY WILL BRING.

1 2 3
4 5 6
7 8 9

Scaling Thickness   Geometrisk Font   *Italics*   Standards

Geometrisk Slim
Geometrisk Light
Geometrisk Regular
Geometrisk Bold
Geometrisk Thick

Geometrisk Slim
Geometrisk Light
Geometrisk Regular
Geometrisk Bold
Geometrisk Thick

Geometrisk   Italics

The Outer Worlds.

Light Italic Slim

Geometrisk Typeface
设计：Daniel Brokstad，BROKSTAD Studio

GEOMETRISK
TYPEFACE
Slim Light Regular Bold Thick
OpenType format   Disk 1/3
For Windows 10 & Mac OS X
Copyright © 2020   Daniel Brokstad

GEOMETRISK TYPEFACE
OTF   v1.0   ©
OpenType format   Release 1.0   Daniel Brokstad
NOW ON SALE!
Windows 10
Mac OS X
Linux
NEW FONTS

WORK HARD PLAY HARD

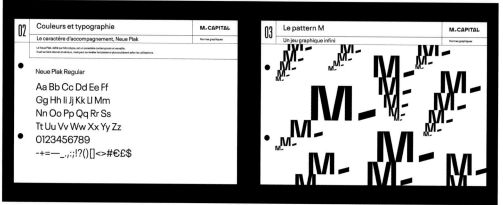

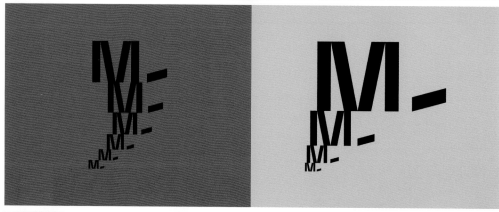

M Capital — Brand identity M Capital 视觉形象设计
设计：Brand Brothers

乌力波人在巴黎

Oulipiens in Paris

Noël Arnaud (writer)/ Michèle Audin (mathematician, writer, former professor)/ Valérie Beaudouin (researcher in social sciences)/ Marcel Bénabou (writer, historian)/ Jacques Bens (writer, poet)/ Claude Berge (mathematician)/ Eduardo Berti (writer, journalist)/ André Blavier (poet)/ Italo Calvino (journalist, short story writer, writer, biographer)/ Bernard (linguist)/ Ross Chambers (professor)/ Stanley Chapman (architect, designer, translator, writer)/ Jacques Duchateau (writer, screenwriter, director)/ Pierre Rosenstiehl (mathematician, researcher)/ Michelle Grangaud (poet)/ Paul Braffort (French computer scientist, engineer, researcher, writer, poet, singer, lyricist, songwriter)/ Marcel Duchamp (painter, sculptor, chess player, writer)/ François Caradec Cerquiglini

Luc Étienne (writer)/ Jacques Roubaud (writer, poet, mathematician)/ Paul Fournel (writer, poet, publisher, cultural ambassador)/ Anne F. Garréta (novelist)

嘉宾 Guest：史亦 Shi Yi (湖岸乌力波系列图书责任编辑)/ Tony Yet (翻译、《二十面体：乌力波》主编)
策划 Curator：二十面体 Icosahedron Magazine
主办机构 Organiser：山中天艺术中心 Wind H Art Center
十点睡觉有限实验室 Click Ten Art Lab

2020/12/12 15:00–17:00

地点 Venue：
山中天艺术中心 Wind H Art Center
北京市朝阳区798艺术区万红里甲31号
31A Wanhongli, 798 Art District, Chaoyang, Beijing

Jacques Jouet (poet, novelist, short story writer, play wright, essayist, visual artist)/ Jean Lescure (writer, poet, screenwriter)/ Étienne Lécroart (cartoonist, illustrator)/ François Le Lionnais (chemical engineer, mathematician, writer)/ Hervé Le Tellier (writer, linguist)/

Albert-Marie Schmidt (linguist)/ Olivier Salon (mathematician, writer)/ Daniel Levin Becker (writer, translator, musical critic)/ Pablo Martín Sánchez (reader, proofreader, translator, writer/ poet)/ Clémentine (translator, novelist, writer, photographer)/ Michèle Métail (choreographer, poet, writer)/ Ian Monk (writer, translator)/ Oskar Pastior (poet, translator, writer, journalist)/ Georges Perec (novelist, filmmaker, poet, documentalist, essayist)/ Raymond Queneau (novelist, poet, critic, editor)/ Raymond Queneau (novelist, poet, critic, film critic/ editor)/ Jean Queval (translator, writer, journalist, film critic)/

Jürgen Ritte (translator, writer, literary critic, editor)/ Frédéric Forte (poet)

Recits d'Ellis Island　Les Lieux d'une fugue

The Video Night of Georges Perec

《埃利斯岛传说》《出走的地点》乔治·佩雷克影像之夜

	No. on List.	NAME IN FULL									

2020/12/19 18:00–20:00

影像作品 Video 1：《埃利斯岛传说》Recits d'Ellis Island
导演 Directors：罗伯特·鲍勃 Robert Bober
乔治·佩雷克 Georges Perec
作品时长 Duration：57分钟 Minutes
作品年份 Release time：1980

影像作品 Video 2：《出走的地点》Les Lieux d'une fugue
导演 Directors：乔治·佩雷克 Georges Perec
作品时长 Duration：40分钟 Minutes
作品年份 Release time：1982

地点 Venue：
山中天艺术中心 Wind H Art Center
北京市朝阳区798艺术区万红里甲31号
31A Wanhongli, 798 Art District, Chaoyang, Beijing

主办机构 Organizer：
山中天艺术中心 Wind H Art Center
十点睡觉有限实验室 Click Ten Art Lab

乌力波 60 周年活动
设计：非白工作室

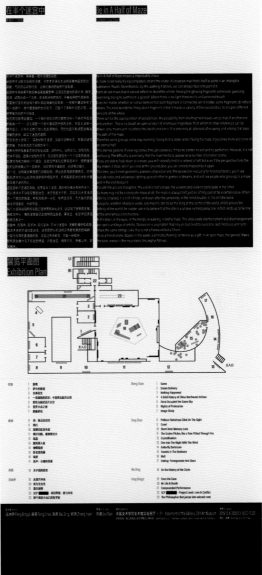

Be in A Half of Maze
设计：非白工作室

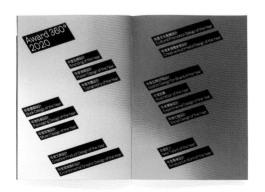

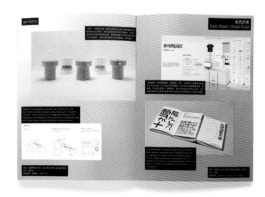

Award360° 2020
设计：非白工作室

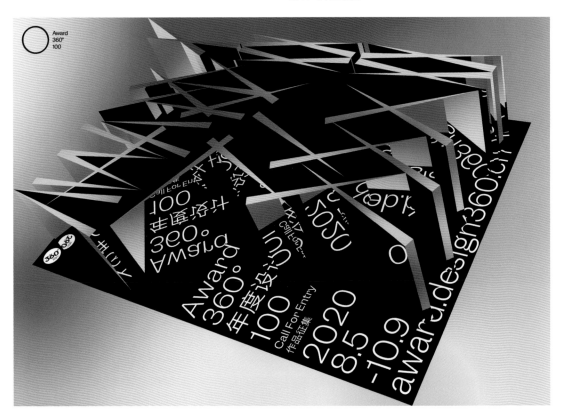

Chinatown Street
[?right Piece of Art?] otos
OS

vas chosen by the New York City Department of [Tra?]
Chinatown Partnership to display this
'Seasonal Street' public design.

Mr. Chen lives in Williamsburg and is from Hangzhou, China. "I felt like I wanted to do something for my people and for the city," he said
Christopher Lee for The New York Times

ade de

ueva York 纽约

"In 201[?] I went
City, a[t] Nom Wa[h]

gothamist.com

worldjournal.com

400 Bad Request
cloudfl[are]

Son

tos: Historic Doyers Street Is
[?]ing A Giant Mural Painted Ove[r]

news.artnet.co[m]

artnet news

4 NEW YORK 66°

Chinatown Street with Bloody Past Now a Bright Piece of Art

Published July 19, 2018 • Updated on July 23, 2018 at 4:37 pm

Dra

一零

Art World Sarah Kolodny

How—and Why—the Artist Chen Dongfan Transformed Chinatown's Doyers Street Into a Flowery, Friendly

Fo

龙与花之歌　The Song of Dragon and Flowers

Public Art Project

艺术家
Chen Dongfan

July - November 2018　2018年7月至11月

Doyers Street, Chinatown, New York　纽约市唐人街多爷街

NYCDOT, Chinatown Partnership　纽约市交通局、华埠改进促进社

Sponsor
Special Thanks

Fou Gallery, Inna Art Space,
[?]ood Printing, Echo He, Inna Xu,
Shelly Hong

Zhongsheng [?], Nadia Lin, Inna Xu

Nadia Lin [?], Inna Xu

Documentary　纪录片

Publisher

Editor

[?] Design

A portion of the images in this monograph are from Instagram. I would like to thank everyone for the excellent photographs they took on the work.

2018

07

06

ON THE ROAD

124
编排设计

Hello everyone! First of all, I would like to thank my wife, Inna Xu, for her care and support. I can only concentrate on my artwork with your companion.

Thanks to the organizers, New York Department of Transportation and Chinatown Partnership. Thank you for choosing me to implement the project, thank you Fou Gallery and Art Bridge to trust me and support me in the creative process.

"Improvise with color and brushstroke, to paint directly on the whole street, turning a 100-year-old street

into a unique piece of art. When people all over the world walk on this road, they can appreciate the com[?] and vitality of Chinatown." When I wrote the artist's statement in the proposal stage, I feel that it is crazy. But the proposal is passed through unanimous vot[e] joined the first meeting remotely via Wechat video ca[ll] as I was still on a holiday trip. When I saw everyone on the screen with passionate faces and serious attitudes, I realized that this is a group of people who are as crazy as me, full of passion and ideals, eager to present the work as soon as possible this summer to New York and to the whole world! Thank you agai[n]

we have achieved a crazy and beautiful work of art together.

Thanks to Echo He from Fou Gallery and Sally Hong from ArtBridge for their persistence and help. Without your recommendation and lobbying I won't decide to submit my proposal and put all my heart in this project- even though I had to end my summer vacation early and miss the four-year World Cup.

I have said in my Facebook and Instagram that only people who have a good heart and beautiful mind will

selflessly dedicate to the project. Thank you all for making the good thing happen!

I want to thank Wellington Chen, Stephen and [?] I would like to thank the friends for their on site [?] behind the stage work, Lin Jing, Fang Yuan, An[?] Dodo Zhang.

I would also like to thank Nadia Lin and Inna Xu[?] their live photography and video recordings.

I would like to thank the Chinatown team for cit[?]

Chen Dongfan Speech at the Unveiling Event of The Song of Dragon and Flowers

and maintenance work every day. I would also like to thank the pioneer [?]

the beginning of a even brighter future.

artwork experiences wind, rain and sun. The color and details of the work will change. The work will amaze

"Love is at the root of everything, all learning [?] repetition all relationships and the core of the lack of

become filled with romance because of love. All these things come together, meet and coalesce 意义也不 because of art. I hope this book can gather together these glorious fragments. Like a family album, flicking through which brings one tender warmth.

To be able to publish this book in 2020, is also of a most uncommon significance. I hope we will soon be able to return to normal.

The Song of Dragon and Flowers
设计：非白工作室

责任编辑：章腊梅

执行编辑：金晓昕

整体设计：成朝晖　祝岩芬

责任校对：杨轩飞

责任印制：张荣胜

图书在版编目（ＣＩＰ）数据

编排设计 / 成朝晖编著. -- 杭州 ： 中国美术学院
出版社，2022.4
　　（平面港）
　　ISBN 978-7-5503-2718-4

　　Ⅰ．①编… Ⅱ．①成… Ⅲ.①版式－设计 Ⅳ.
①TS881

　　中国版本图书馆CIP数据核字(2021)第216042号

平面港
编排设计
成朝晖 编著

出 品 人：祝平凡
出版发行：中国美术学院出版社
地　　址：中国·杭州南山路218号　邮政编码：310002
网　　址：http://www.caapress.com
经　　销：全国新华书店
印　　刷：杭州恒力通印务有限公司
版　　次：2022年4月第1版
印　　次：2022年4月第1次印刷
印　　张：8
开　　本：787mm×1092mm 1/16
字　　数：150千
印　　数：0001—3000
书　　号：ISBN 978-7-5503-2718-4
定　　价：38.00元